All the Secrets of the Solar System

In Large Print

Harold A. Geller

Photo Credits:
Front Cover – NASA/JPL
Back Cover – S.M. Zoltek

ISBN 978-1-304-34530-1

Table of Contents

Chapter 4
Light, Optics, and Telescopes

Chapter 5
Matter and the Study of Radiation

Chapter 6
The Earth and Moon

Chapter 7
The Terrestrial Planets

Chapter 8
The Jovian Planets

Chapter 9
Minor Bodies and the Interplanetary Medium

Chapter 10
Formation of the Solar System

Chapter 11
Radiation from the Sun

Chapter 12
The Outer Layers of the Sun

Chapter 13
The Quest for Life

Chapter 14
Solar System Revelations

DEDICATION

I dedicate this book to the memory of my late brother, Richard B. Geller (1946-2011). He dedicated his life's work to the education of students, especially in mathematics. Richard taught in the New York City public school system for over forty years, the last thirty years at Stuyvesant High School. May his mantra, "Math is #1," keep the memory of his work alive, and may the readers of this book seek to understand the secrets of the Solar System as Richard sought to teach the secrets of this Universe through his beloved mathematics.

ACKNOWLEDGEMENT

I gratefully acknowledge John C. Evans, Emeritus Professor of Physics and Astronomy at George Mason University for his contribution to this volume. My former colleague provided me with an unfinished original manuscript upon which this book is based. I have copied, pasted, added, deleted, modified and edited the original manuscript with his expressed permission and blessing. If there are errors, the fault is mine, and if it provides the reader with knowledge and understanding of our Solar System, the credit should go to Dr. John C. Evans. It is hoped that this effort brings to its audience, especially those with some impaired vision, all the secrets of the Solar System.

Chapter 1
Secrets of the Solar System

All that we know about our solar system today is far different from that held a few thousand or even a few hundred years ago. In fact, our view of the entire universe continues to change. The universe, as we define it, refers not only to the celestial phenomena with which we are acquainted now, but also to whatever new ones may yet be discovered. The ancient Greeks referred to the universe as the cosmos. This was their word for an orderly and harmonious system.

Astronomy is a science that has captured popular imagination. Many view astronomy as a science synonymous with the future. However, astronomy may well be the oldest of the sciences or perhaps even the "mother" of all science. The human pursuit of knowledge about the physical world has been strongly shaped by our fascination with the changes that we see in the heavens. Intellectual thought has been inspired by both the beauty and the immensity of the astronomical world. Astronomy has not only stimulated other sciences, such as physics and mathematics, but it has also inspired creative efforts in art, music, and literature. Let us begin our exploration of the cosmos by seeking the origin of science itself.

1.1. HISTORICAL ORIGINS OF SCIENCE

History shows us that one of the human preoccupations has been with the nature of our physical existence. What are we made of? How did we come into being? What is our fate?

No other experience, outside of human physiological functions, is any more universal to the peoples of the Earth than that of the heavens. We can trace the roots of this concern about our physical existence back thousands or even tens of thousands of years when our ancestors first began contemplating the night time sky. Cave paintings, cliff carvings, stone and wood monuments, and pottery fragments found worldwide all testify to the importance that the sky has played in early thought. For example, we have evidence that stone-age humans in Europe were keeping track of the cycles of lunar phases on pieces of bone some 30,000 years ago.

To many of our ancestors the sky was more than a source of wonder. In fact, the sky above appeared to hold power over our earthly existence. Celestial gods were believed to be able to control human destiny. Astrology, which is not science, was believe to be the key that revealed divine plans for the course of human events. Without any scientific basis and having clearly diverged from astronomy by the seventeenth century, early astrology did make at least one contribution to astronomy. Astrologers recorded the regularities in the movement of the Sun, Moon, and planets relative to the background stars. Because of this, well before the dawn of recorded history some 6000 years ago, these regularities were part of the human knowledge base. Some of the earliest astronomical texts are found on Babylonian cuneiform tablets dating from almost 2000 B.C. They record the movement of the planets. The word planet itself comes from the Greek language, and it means "wanderer," in contrast with the fixed stars.

Concern for astronomical knowledge was not confined to the civilizations of the Mediterranean. Some time in the third millennium B.C., before the building of the pyramids in Egypt, early Britons began the construction of Stonehenge. The orientation and structure of Stonehenge has long been known to have astronomical significance. The principle axis of Stonehenge marks the direction of sunrise at the time of the summer solstice. Hundreds of stone monuments in Scotland, England, and France have also been shown to display knowledge of astronomy far beyond that expected of the early inhabitants of these areas. Similar comments can be made about the astronomical significance of many New World structures, such as Carocol in Mexico.

Most ancient peoples devised an explanation for what we can loosely call the origin, structure, and evolution of the universe. This is the subject matter of modern cosmology. Most civilizations believed that out of a great dark void or chaos, the world was created by divine intervention. In these cosmologies we find a common thread. That is, an inclination for people to believe their known world to be the center of the universe. This perspective blossomed into a geocentric model of the universe. That is, the whole Earth was at the center. Ancient cosmologies were only idealized sketches against which the activities of nature took place. Few, if any, aspects of natural phenomena were actually incorporated into these ancient cosmologies. Not until many centuries later were explanations of the details of nature considered to be a necessary part of science and in particular cosmology.

The important question is whether or not these concerns of various early civilizations are primitive beginnings on the road that ultimately led to modern science. In spite of the apparent existence of this rudimentary scientific thinking by various peoples worldwide, scholars strive to understand how all of these diverse activities eventually merged to produce the science we recognize and practice today. We know that much of the purpose behind the acquisition of astronomical knowledge was to produce calendars for agricultural purposes and navigational guides for seafaring nations. Such purposes are an outgrowth of the survival instinct. What we must do is to differentiate between survival activities which are on the road leading to modern technology and those activities that are an inquiry into the nature of the universe. Since the earliest of times, there has existed in humans a theological-philosophical strain concerned more with the nature of our existence than simply survival as a species. It is possibly not unreasonable to say that science is an outgrowth of such theological-philosophical motivations. Current scholarship suggests that Greek philosophy alone is the predecessor of modern science. To most scholars it seems an inescapable conclusion that the Greeks were the originators of that method of thinking known as science.

According to Aristotle (384-322 B.C.), the dawn of systematic scientific thinking began in the sixth century B.C. in the Hellenic cities of Ionia in western Asia Minor. The times were those following the Homeric period (900-700 B.C.) when the eastern end of the Mediterranean was in great upheaval because of the invasion and destruction of the highly developed civilizations of Knossos, Mycenae, Pylas, and others. This era can be compared to that following the re-emergence that took place in Europe centuries later after the collapse of the Roman Empire. The Ionian cities were prosperous and involved in wide-ranging commerce. Unfortunately little remains of their written texts from that period. What we have are commentaries by later writers, such as Aristotle, of the philosophical activities that began in Ionia. Even though we have only fragments of their work or hearsay reports concerning these pre-Socratic philosophers, enough of Greek philosophy of the fifth and sixth centuries B.C. has been passed on so that various themes can be traced.

The first Ionian philosopher of whom anything is known was Thales (632-546 B.C.) of Miletus. He said that water is the fundamental substance and all things are derived from it. Exactly what he meant by this we do not know, as no written record by him survived to this era. Aristotle is our authority on Thales and he seems uncertain himself as to Thales' meaning. We are left to guess and to surmise that what he was proposing was the concept of a unity that permeates nature. What did Thales observe that lead him to propose such a startling idea? Was it an observation of the cycle of water falling as rain, collecting in rivers to run to the sea, there to evaporate forming clouds to fall again as rain? Or did he observe the intimate connection between biological processes in living matter and water? We shall probably never know for sure. But lacking evidence to the contrary, scholars are persuaded that the thoughts of Thales are as good as any in which to place the origin of science.

More is known of Anaximander (about 590 B.C.), a somewhat younger Milesian. In his writings, we find a fundamental theme found in later Greek thought. He imagines the cause of things not in a mystical or mythical way. Unlike Thales hypothesis that a fundamental substance like water is the source of unity in the physical world, Anaximander postulates that a featureless matrix, called "the Unlimited" or "the Infinite," is the source of physical existence by a separation of opposites. Exactly what he means by this we are not sure. Although his world system is not rooted in mechanism as we might argue today, neither is it rooted in mysticism as his predecessors contended. Its roots are in law: All natural processes, he wrote, are governed by an overriding principle of cosmic justice, or Necessity. By denying man's preferred status in nature, he asserts that things happen because they must, which was the first step on the road to scientific rationalism.

At this point we need to leave Greek natural philosophy to consider those astronomical concepts and observations that were the foundation for the two competing Greek geometrical theories of the nature of the world, the geocentric and the heliocentric concepts.

1.2 ORGANIZATION OF THE SKY

In the clear skies of the Tigris, Euphrates, and Nile valleys, where the earliest civilizations flourished more than 5000 years ago, watchers of the heavens singled out and named various groupings of stars, called constellations, primarily for calendrical and navigational purposes. To aid their memory, they imagined that they saw in these groupings the likenesses of mythological beings, animals, and monsters and named the constellations accordingly. The names and shapes of constellations are part of our heritage from ancient Greece, who in turn inherited them from these older civilizations. Greek astronomers identified and named 48 constellations. Forty more were added, most of them in southern skies, by European mapmakers and astronomers in the seventeenth and eighteenth centuries.

Standing in mid-latitudes of either hemisphere a practiced observer can see about four-fifths of the constellations during the course of the year. Star maps containing the constellations for each of the four seasons can be found inside the front and back covers to help you. Of the 88 constellations, about half lie in the Milky Way or near its borders. As you learn the constellations, hearing the name of one will bring to mind an area of the sky, just as earthbound place names identify a particular geographical area.

The "catch figure," or asterism, often associated with a constellation should not be assumed to be the outline of the constellation's namesake. An example is the asterism of the Big Dipper, which is the recognizable figure for the constellation Ursa Major, the Great Bear. In today's astronomy, the constellations define specific areas of the sky with north-south and east-west boundaries. All celestial objects lie within the borders of one of the 88 constellations. Stars that identify a constellation form an apparent grouping as seen from Earth and are not necessarily in proximity to each other in space.

To ancient peoples, the sky appeared to be the inside of an immense dome covering the Earth and extending as far as they could see. Little wonder that they considered the Earth as a small sphere, or even a flat plate, at the center of a very huge sphere. And the stars that they saw were considered to be located just inside or on the large sphere. This two-sphere concept is still used by astronomers for organizing the sky. The imaginary outer sphere is known as the celestial sphere.

Watching the sky, early peoples could see that stars rise above the eastern horizon, the circle that divides the visible celestial hemisphere from the invisible one, cross the sky during the night, and later set below the western horizon. This daily, or diurnal, behavior was attributed by most peoples, including the Greeks, to the rotation of the celestial sphere from east to west. Today we know that the apparent rising and setting of stars is actually caused by the Earth's rotation in the opposite sense, that is, from west to east.

Ancient peoples also observed that during their diurnal motion the stars move around two points on the celestial sphere, the north and south celestial poles or, NCP and SCP for short. To them these were the ends of the axis about which the celestial sphere rotated. To us they are the points of intersection of the Earth's axis of rotation with the celestial sphere. The significance of Polaris, the North Star, is that it lies within $1°$ of the north celestial pole and is a relatively bright marker of the NCP's position.

For any observer, the point directly

overhead is called the zenith. And the imaginary arc on the celestial sphere running from the north point of the horizon through the north celestial pole and zenith to the south point of the horizon is called the celestial meridian. This line is the dividing line between rising and setting, since the highest point above the horizon that any star reaches in its daily motion occurs when the star crosses the celestial meridian. For this reason, the celestial meridian is a basic reference in timekeeping.

Which stars are or are not visible depends on the observer's geographic latitude. An observer in northern latitudes notices that not all stars rise above and set below the horizon daily. Stars near the NCP are always above the horizon, while stars near the SCP are always below the horizon. As ancient peoples traveled to different latitudes, they noticed that the stars that were visible to them changed. For example, as travelers moved northward, they could see stars near the northern horizon that previously had not been visible, and stars previously visible near the southern horizon were now below it. Such an effect along with some others was used as evidence that the Earth is spherical and not flat.

Over the course of the year, the Sun does not rise due east and set due west; only twice during the year will it actually do this. During the rest of the year and depending on the observer's latitude, the Sun shifts its rising and setting position relative to the east and west points.

At the time of the summer solstice, the first day of summer in the Northern Hemisphere which occurs on or about June 21, the Sun rises as far north of the east point and sets as far north of the west point as at any time during the year. Over the three months of summer, the Sun moves back south toward the east and west points when it rises and sets. During summer, the hours of daylight are longest and the Sun is highest in the sky. At the time of the autumnal equinox, the first day of autumn, which occurs on or about September 22, the Sun rises due east and sets due west. On this day, all places on Earth experience 12 hours of daylight and 12 hours of darkness. During the three months following the autumnal equinox, the Sun continues its southward migration in the sky, so that by the time of the winter solstice, the first day of winter, which occurs on or about December 22, the rising and setting points are as far south as at any time during the year. At the time of the winter solstice, the angular distance of the rising and setting points of the Sun south of the east and west points is equal to the angular distance of the rising and setting points north of the east and west points at the time of the summer solstice. The length of daylight is now the shortest, and the Sun is lowest in the sky. Over the six month period from winter solstice through the time of the vernal equinox, the first day of spring, which occurs on or about March 21, to the summer solstice, the Sun rises and sets farther north each day. At the time of the vernal equinox, the Sun rises due east and sets due west, when there are again 12 hours of daylight and 12 hours of darkness.

1.3. CYCLIC PHENOMENA OF THE HEAVENS

Historically, the common cyclic phenomena of the sky, such as the Sun's daily rising and setting, played an important role in the development of the geocentric and heliocentric concepts. Many of these phenomena were not discovered in any literal sense, but had been observed and known long before anyone wrote about them. What is new and has changed over time is an explanation in either the geocentric or heliocentric conceptual scheme of why, for example, the Sun rises and sets. Rather than trace their historical development, let us consider only our present understanding of these cyclic phenomena.

The different seasons are caused by the tilt of the Earth's axis of rotation relative to its orbital plane and by the Earth's revolution once a year about the Sun. As seen from Earth, the Sun appears to move eastward about 1° each day relative to the stars. Because of this, stars rise about 4 minutes earlier each night than they did the previous night. Consequently, by the end of a month, stars are rising approximately 2 hours earlier than they did in the previous month. At the end of a year, the nightly change adds up to 24 hours, and the annual cycle of the heavens begins again.

Over time most ancient peoples were able to identify the Sun's yearly path through the constellations, a path called the ecliptic. Earth's geographic equator projected onto the celestial sphere produces an imaginary line called the celestial equator. The celestial equator intersects the ecliptic at two points, the vernal equinox and the autumnal equinox. The Moon and planets move almost entirely along a narrow band of sky, the zodiac, which is 16° wide and is centered on the ecliptic. The zodiac is divided into 12 constellation divisions, or signs, through which the Sun passes in successive months.

The Earth's axis of rotation is tilted 23.5° to its orbital plane, which is consequently the angle between the ecliptic and the celestial equator. In the Northern Hemisphere we incline away from the Sun in December and toward it in June. Consequently, the amount of sunlight falling on the surface of either hemisphere varies depending on whether the hemisphere is inclined toward or away from the Sun.

In the Northern Hemisphere spring begins in March at the time of the vernal equinox, when the Sun crosses the celestial equator from south to north. Summer starts in June at the time of the summer solstice, when the Sun reaches its maximum distance of 23.5° north of the celestial equator. The next season, autumn, begins in September at the time of the autumnal equinox, when the Sun crosses the celestial equator from north to south. Finally, winter commences in December, when the Sun reaches its maximum angular distance of 23.5° south of the celestial equator at the time of the winter solstice. In the Southern Hemisphere, the seasons are reversed; for example, Christmas occurs there during the warm summer months.

The daily rising and setting of the Sun, the year of the Sun's seasons, and the monthly period for the phases of the Moon were important cyclic events for ancient peoples. All these repetitive cycles were important to them in establishing a concept of time.

Our understanding of the reasons for the Moon's phases predates even Aristotle, who was aware that the Moon "shines" by reflecting sunlight. The parallel rays of the distant Sun always illuminate one-half the Moon's surface as well as one hemisphere of the Earth. When the Moon is visible, we are seeing rays of sunlight reflected off the Moon's surface.

When the Moon is between the Earth and the Sun, its dark side faces us, and we do not see it. This phase is called New Moon. Because the Moon moves eastward relative to both the Sun and stars, within a few days a thin crescent appears low in the western sky after sunset and sets shortly after the Sun. In the next few days, a growing crescent appears higher in the sky after sunset and therefore sets later on successive nights. One week after New Moon, the Moon is at first quarter and will be on the celestial meridian at sunset. It will set about six hours after the Sun. In the following week, the Moon becomes full as it continues its easterly movement around the Earth. Two weeks after new moon is the time of full moon, when the Moon lies on the opposite side of the Earth from the Sun. We see it rise at approximately 6 P.M. and set about 6 A.M. One week later the Moon is at last quarter where it rises about midnight and sets around noon. Finally, we see the declining crescent-shaped Moon rise shortly before sunup as it is about to overtake the Sun one month after the previous new moon.

If we observe the Moon's movement against the stars, we find that it moves a little over a $0.5°$ per hour, or about $13°$ per day. Thus the Moon takes around 27.3 days to complete its orbit of $360°$; this period is called the sidereal month. However, because the Earth is also moving around the Sun, the time between two successive cycles of lunar phases is longer than the sidereal month. Although the Moon has completed its revolution around the Earth at the end of 27.3 days, it takes about two more days to bring the Moon back to the Earth-Sun line so that it again appears as a new moon. Thus the period of lunar phases, the synodic month, is 29.5 days.

Although the Moon is some 400 times smaller than the Sun, it is also some 400 times closer to us, meaning that both the Sun and Moon have the same angular size, about $0.5°$, as seen from the Earth. This produces one of the most impressive and yet important scientific coincidences in nature: When the Moon comes directly between the Earth and the Sun, the Moon's shadow or part of it falls on the Earth's surface. As seen from that point the Moon covers the Sun, blocking out the Sun's light. This is known as a solar eclipse.

A solar eclipse is possible only when the Moon is near a new phase. Since the plane of the Moon's orbit is inclined to the Earth's orbital plane by about $5°$, the Moon must additionally be at or near one of the two points in its orbit where that orbit intersects the Earth's orbital plane. This lineup occurs at least twice each year and at most, but rarely, five times a year.

The totally dark portion of the shadow the Moon casts during a solar eclipse is called the umbra; the penumbra is the partial shadow or semi-dark portion. If one stands in the umbral shadow, one sees the Sun completely covered by the Moon, and this is called a total solar eclipse. If one stands in the penumbral shadow, one sees a partially covered Sun, and this is called a partial solar eclipse. The penumbral shadow covers a much larger area on the Earth's surface than does the umbral shadow, making a partial solar eclipse visible over a wider region than a total solar eclipse.

The average length of the Moon's conical shadow is not quite equal to the Moon's mean distance from the Earth. An annular solar eclipse takes place when the Moon, in addition to the conditions under which a partial or total solar eclipse occurs, is also farthest from the Earth. At this point its umbral shadow is too short to reach the Earth, so that one sees the slightly smaller, darkened disk of the Moon surrounded by a brilliant ring of still-exposed Sun.

Under the most favorable conditions in the Earth's equatorial regions, the Moon's umbral shadow is some 270 km wide, while the penumbral shadow is close to 6000 km wide. At this time, the totality of the eclipse lasts longest along the path of the eastward-moving shadow, the maximum length being about 7.5 minutes.

Usually, if an eclipse of the Sun occurs, an eclipse of the Moon precedes or follows it by two weeks. The Earth, Moon, and Sun are then sufficiently in line for the full Moon to move totally or partially into the Earth's shadow producing a lunar eclipse. Since the Earth's diameter is nearly four times that of the Moon, the conical-shaped shadow cast by the Earth is about four times wider at the base and four times longer than the Moon's shadow. Lunar eclipses may be partial or total, everyone on the dark side of the Earth seeing the lunar eclipse at the same time.

A year may bring as many as three lunar eclipses, or none at all. More often we have two eclipses of the Sun and two of the Moon in each calendar year. Centuries of observing eclipses taught the Babylonians that eclipses recur at regular intervals. After 18 years and 10 or 11 days, the circumstances of an eclipse are repeated approximately. By 200 B.C., Babylonian astronomers could predict with surprising accuracy future lunar eclipses. Their prediction method came about by noting numerical relations, what we today would call a numerical algorithm, in tabulated observations, rather than devising a geometrical relationship for the Sun, Moon, and the Earth as the Greeks later did.

Ancient astronomers devised names to identify particular positions of the planets relative to the Earth and Sun on the celestial sphere. This early system forms the basis for current definitions of planetary configurations. Between the Earth and Sun revolve Mercury and Venus; because of their smaller orbits they are called the inferior planets. As seen from the Earth and measured relative to the Earth-Sun line, Mercury and Venus appear to move counterclockwise around the Sun while swinging from one side of the Sun to the other. The angular distance, in degrees, that any planet appears east or west of the Sun is called its elongation.

From a position closest to the Earth called inferior conjunction, when it is in line with the Sun, either of the inferior planets appears to move rapidly westward from the Sun, which causes its phase to change from new to crescent. When the inferior planet reaches its greatest angular distance west of the Sun, known as maximum western elongation, it is conspicuous in our skies as a "morning star" and its phase is quarter. Thereafter, the planet appears to reverse its course and move eastward back toward the Sun until its elongation is a minimum, a configuration known as superior conjunction. At this point, the planet is on the opposite side of the Sun from the Earth, and its phase is full. Leaving superior conjunction, an inferior planet continues to move eastward on its way toward its greatest angular distance east of the Sun, known as maximum eastern elongation. Here it is seen in the heavens as an "evening star" and its phase is also quarter. Next, the inferior planet moves back toward the Sun and inferior conjunction, completing its cycle of planetary configurations and moonlike phases.

Planets with orbits outside Earth's are called superior planets. To the ancients, the superior planets were the three naked-eye planets Mars, Jupiter, and Saturn. When a superior planet is nearest to us and also brightest, its configuration is known as opposition, opposite the Sun in the sky and visible throughout the night. Although the superior planet moves eastward relative to the stars, it lags behind the swifter Earth's motion and so appears to drift westward from the Earth-Sun line until it is 90° east of the Sun, at which point it is in eastern quadrature. Here it rises at noon and is an "evening star." Although north and south are absolute directions in space defined by the Earth's axis of rotation, east and west are a sense of rotation defined by the Earth's rotation from west to east. Continuing to drift eastward relative to the Earth-Sun line, the superior planet's elongation decreases; it is thus approaching the Sun. It arrives at a configuration known as conjunction, where the superior planet is on the opposite side of the Sun from Earth and will rise and set with the Sun. From here the superior planet passes through western quadrature, 90° west of the Sun. At this point the superior planet rises at midnight and is a "morning star." Finally, the superior planet returns to opposition, its cycle of configurations complete. As seen from the Earth, superior planets do not exhibit a cycle of moonlike phases as do the inferior planets.

The length of time for one orbit of a planet around the Sun is known as its sidereal period. It is the time taken to complete a 360° circuit around the sky relative to the stars. Unfortunately, there is no marker along a planet's orbit to indicate when it has completed a 360° revolution. From the Earth we actually observe what is called the synodic period, that is, the time it takes a planet to return to a particular configuration with respect to the Earth-Sun line, such as from opposition to opposition. The synodic and sidereal periods differ because the Earth is advancing in its own orbit as a given planet revolves around the Sun. A simple mathematical relation, known since the time of the Greeks, allows us to calculate the sidereal period after measuring the synodic period.

Since Earth's orbital period is shorter than that of a superior planet, the Earth overtakes a superior planet and passes it. This occurs while the planet's configuration changes from western quadrature through opposition to eastern quadrature. During this period of passing, the planet appears to temporarily interrupt its normal eastward motion relative to the stars and move westward. This countermotion is known as retrograde motion, in which the superior planet executes a closed or open loop and then continues its usual path eastward relative to the stars. Relative to the Earth-

Sun line, however, it is moving toward an area of the sky east of the Sun.

Now having surveyed the common Earth-Sky relationships and the regularity of the heavens, and remembering that the Greeks and other peoples of that period were well aware of these regularities, we can discuss the historical development of astronomy, leading up to the modern conception of the dynamics of planetary motion.

1.4. GREEK COSMOLOGY

Although there is a tendency to dwell on those ancients who may have envisioned the Earth as flat, knowledge that the Earth and Moon are spherical was widespread in the Greek world by the fifth century B.C. Aristotle argued that the circular shadow projected by the Earth when it eclipsed the Moon was clear evidence of the Earth's spherical nature. This argument had been known for a long time.

As Hellenistic culture spread throughout the eastern Mediterranean world, a new establishment for science was centered in

Alexandria after about 300 B.C. The Museum and its associated Library in Alexandria was one of the most famous centers of learning in the ancient world. The Museum was a center for scientific and mathematical research. One of its geographers, Eratosthenes (273-193 B.C.), knew that the Earth was a sphere, and from the earlier works of Aristarchus (310-230 B.C.), that the Sun was at least 20 times farther away than the Moon, although the correct value is nearer 400. Eratosthenes reasoned that rays of sunlight ought to be parallel when they reach the Earth, enabling him to measure the Earth's circumference by the geometrical argument.

Eratosthenes chose observing stations at Alexandria and Syene to the south, where the Aswan Dam is now located on the Nile River. For the time of the experiment he chose local noon on the day of the summer solstice, which comes at the same moment at both sites because they are very nearly on the same meridian of longitude. He probably selected that day, since the Sun was as far north as it would be during the year, meaning that it would pass very near the zenith at local noon at Syene.

At noon an observer in Syene observed that the Sun was directly overhead, while an observer in Alexandria found the Sun to be 7° south of the zenith. Measurers had paced off the distance between the two cities as about 4900 stadia, where one stadium is estimated to be approximately 0.16 kilometer. Because a straight line cuts two parallel lines at equal angles, the angle at the center of the Earth is equal to the zenith angle of 7°. Working a simple proportion, Eratosthenes calculated the Earth's circumference as follows:

$$C/4900 \text{ stadia} = 360°/7°$$

Thus the circumference of the Earth was discovered to be 252,000 stadia, or about 40,320 kilometers (km). In principle, Eratosthenes' experiment was correct. Although his measuring technique was inaccurate by modern standards, his results were surprisingly close to today's mean value of 40,030 km.

As early as 2000 B.C. the Babylonians had begun recording the movements of planets, and a thousand years later the Greeks had inherited much of that knowledge. The Greeks, who to a great extent are responsible for developing geometry and trigonometry, sought a geometrical explanation of planetary motions rather than the simply numerical relationships found by the Babylonians. For them to see the Earth, which was not in their minds a planet, as the center of a world system, was not unreasonable, even though they were aware of the difficult concept of a spherical Earth. A geocentric, or Earth-centered, cosmology certainly seems more intuitively obvious than does a heliocentric, or Sun-centered cosmology. For centuries, various Greek schools of philosophy proposed, debated, and elaborated several geocentric theories.

Any conceptual scheme had to explain the observed motions of the Sun, Moon, and five naked-eye planets, all of which seemed to wander among the stars. Never changing their course, the Sun and Moon move eastward on the celestial sphere in a somewhat steady fashion relative to the stars. The inferior planets, however, move eastward through the background stars of the zodiac for a time and then undergo retrograde motion, moving westward. The superior planets generally follow an eastward path relative to the background stars with only a brief period of retrograde motion westward. All the planets further confound an understanding of their motion by moving swiftly some times and slowly at other times while displaying noticeable variations in brightness.

Greek philosophers, with their aesthetic taste for symmetry, reasoned that nature arrayed her celestial bodies on the perfect geometric figure, the sphere, and moved them in the perfectly symmetric plane figure, the circle. Beginning with Plato and possibly earlier, generations of astronomers thought that planetary movements must be accounted for by combinations of uniform circular motions with the Earth at the center. With this geocentric concept, they tried a variety of conceptual models to account for the observed motions. The ultimate product of geocentric cosmology was the Ptolemaic system.

The heliocentric concept, which followed the geocentric one, did not originate with Copernicus. He became aware that in the third century B.C. the Greek natural philosopher Aristarchus had proposed the Sun as the center of planetary motion. In his treatise, On the Sizes and Distances of the Sun and Moon, Aristarchus estimated that the Sun is 20 or so times farther from the Earth than the Moon, although the actual value is about 400, and since both have approximately the same angular size, the Sun must be 20 times larger than the Moon. Thus he reasoned that the Sun was about 7 times the Earth's diameter, although the actual value is almost 109 times. From these estimates he apparently thought it natural to put the largest and only self-luminous body in the Solar System, the Sun, at the center of the system. Additionally, Aristarchus attributed the daily movement of the heavens to the rotation of the Earth on its axis. Annual changes in the sky and the planet's motions could be explained if they and the Earth then revolved about the Sun. Even prior to Aristarchus, the Greeks were aware that the Moon, and possibly the planets, "shine" by reflecting sunlight, a notion they probably came to by observing lunar eclipses. It is also possible that Aristarchus recognized that the stars were self-luminous and conceivably like the Sun only farther away, but this is speculation.

Aristarchus's contemporaries objected to the heliocentric concept for several reasons: If the Earth does move, why don't we feel its motion, which must be exceedingly fast? Moreover, if the Earth moves, why don't we see every unattached object on the Earth's surface sailing swiftly past us in the opposite direction? However, what seems to have been the strongest argument against the heliocentric concept was the failure to see a shift (which would be due to the Earth's orbital motion) in the apparent position of nearby stars relative to more distant stars, a phenomenon known as parallax. We know today that nearby stars do appear to change place, but they are so far away from us that the apparent angular displacement, or parallax, is extremely small, even for the closest stars, and can be measured only with a large telescope.

The fact that the parallax phenomenon could be adduced as a criticism suggests that the Greeks possibly were aware that the stars are not all at the same distance from the Earth, a point they may have arrived at by arguing that all the stars have about the same brightness and that therefore the fainter ones are the more distant ones. However, even though they may have suspected that the stars are not all at the same distance from us, they apparently did not realize that even the nearest stars are too incredibly distant to reveal any parallax to the naked eye.

Within a couple of centuries after Aristarchus, the heliocentric cosmology had lost out to the geocentric until its revival by Copernicus. Several generations of astronomers at the Museum in Alexandria set for themselves the goal of removing discrepancies between the geocentric concept and observed planetary motions. They felt that their revised geocentric concept must retain the circular motions and uniform rates appearing in such theories as Aristotle's geocentric system. They fashioned their geocentric model on the basis of several earlier natural philosophers, especially Heraclides (390-310 B.C.), Apollonius (262-190 B.C.), Hipparchus

(190-120 B.C.), and a later one of their own, Ptolemy (A.D. 90-168). The resulting system had a number of combinations of circles and off-center motions. To be an acceptable conceptual scheme, the geocentric theory had to represent more accurately the many small cyclic changes and large general motions of the world system. In this they succeeded, and the final version, which dominated philosophical thought for 13 centuries until the time of Copernicus, was published around A.D. 150 in the Syntaxis of Astronomy, an astronomical encyclopedia compiled by the last of the great Alexandrian astronomers Claudius Ptolemy.

Ptolemy's geocentric system, taken in part from the earlier work of Heraclides and Apollonius, presented each planet as moving uniformly around a small circle called an epicycle. The center of the epicycle in turn revolved uniformly around the circumference of a large circle called a deferent. By means of proper combinations of sizes and rates of motion for the epicycle and deferent, planetary motions could be mostly direct and occasionally retrograde.

Also, since a planet on an epicycle is sometimes nearer and sometimes farther from the Earth, this accounted for the observed variations in planetary brightness. To represent the irregular rates of motion of the planets, Ptolemy continued to employ the device attributed to Hipparchus of having the deferent off center from the Earth (to produce an eccentric deferent), so that a planet would appear to go fastest when it was closest to the Earth.

Having constructed orbits for the Sun, Moon and planets out of a combination of epicycles and eccentric deferents, Ptolemy found that the heavenly bodies were moving at an even more irregular rate than could be accounted for by these devices. His solution to this problem was to suppose that the planets' motions were uniform not as viewed from either the Earth or even the center of the eccentric deferent, but from a point on the other side of the center of the deferent from the Earth; this point was called the equant.

Why was science invented by the Greeks and not the Babylonians, the Egyptians, or any other peoples? Evidence to answer such a question has long since vanished, if it ever existed, and we can but speculate. We do know, however, that the Greeks, although bound by common cultural ties, were not organized into a single, rigid, monolithic state as were many other peoples at that time. Greece was a loose confederation of self-governing city-states. Besides science the Greeks gave the world both democracy and a language capable of expressing subtle distinctions in concepts. Thus the Greeks perceived not only a physical world operating under laws, but they conceived of a rule of law governing social order and incorporated that vision in to their very language. Whether one or the other of these three contributions, language, democracy, or science, led to the others, or all three are by-products of some interaction or conditioning by their peculiar environment, we can not say. But these three elements of what the mathematician and science commentator Jacob Bronowski (1908-1974) called the ascent of man, go hand-in-hand, and tell us much about the mental and emotional outlook of the ancient Greeks.

Ptolemy's cosmological model lasted until Copernicus challenged it in the sixteenth century when he declared that the Earth is a planet and the Sun is the rightful occupant of the center of the Universe. Even then, however, Ptolemy's system did not totally disappear until the time of Newton in the late seventeenth century. But the belief that the Earth, the Sun, the Moon, and the planets must occupy the center of the Universe, since man is apparently the center of creation, did not completely disappear until the beginning of our own century, almost 350 years after Copernicus's death.

Chapter 2
Motions in the Heavens

Almost 3000 years after the Babylonians began recording motions in the heavens and 1300 years after Ptolemy's geometrical explanation of those motions, Nicolaus Copernicus revived Aristarchus's heliocentric concept launching astronomy, and consequently science, on a new road to understanding motion. On Copernicus's heels followed Tycho Brahe with new and better observations of the motions of planets and Johannes Kepler with his ingenious laws describing planetary motion. Briefly stated, Kepler's first two laws are that the orbits of the planets are ellipses, not circles as Plato had wanted, and their variable motion is due to their changing distances from the Sun. His third law establishes the relationship between the planets' orbital periods and their mean distances from the Sun. Kepler's laws are universal in that they apply to any two bodies gravitationally bound to each other, whether in the Solar System or elsewhere in the Universe.

Shortly after Copernicus and Kepler

came a fascinating personality and the architect of a new perception of motion in the person of Galileo Galilei. The 200-year period spanned by Copernicus, Kepler, and Galileo can be thought of as a dividing line between an old science, which had seen only minor changes since its invention by the Greeks, and a new science in which the methods of observation and experimentation would play an increasingly more profound role. There is, however, in this transition period no abandonment of the basic elements of scientific thinking as first laid down by the Greeks. Therefore let us begin this chapter by considering scientific concepts, elements on which physical laws and theories are based.

2.1. PHYSICAL CONCEPTS IN SCIENCE

In the previous chapter, several concepts, such as the "geocentric" concept or the concept of "a spherical Earth", were mentioned. Later we will encounter additional concepts such as the concept of "the electron," the concept of "the photon," the concept of "force," or the concept of "energy." Let us see if we can determine what is implied by such phrases. In doing so, our discussion of concepts, models, laws, theories-elements in the processes of science-in this and following chapters has been heavily shaped by Gerald Holton's and Stephan Brush's book, Introduction to Concepts and Theories in Physical Science. Their book contains many references for additional reading on the conceptual foundations of physical science.

It is clear historically that scientists limit themselves to certain types of observations and thought processes. Their use of concepts as a reference frame for understanding, their rules for fashioning conceptual schemes, and their type of argument for seeking agreement are unlike those of non-science. Science has accumulated since the time of the ancient Greeks a set of internationally accepted and reasonably enduring concepts, which are its medium of exchange. To a large degree the secret of science's successful harmony and continuity lies in the nature of these concepts, their definitions, and their ready acceptance as the "coin of the realm."

To make concepts more definite, let us adopt for our discussions the following definition of a physical concept taken in part from Holton. A physical concept is a general idea, or notion, or understanding as derived from specific instances or occurrences in the natural world. In general, physical concepts can be defined operationally and most possess a quantitative nature.

The characteristics of concepts-possessing operational definitions and quantitative natures-actually do more to define the notion than does the first sentence. Each of these characteristics aid in the unambiguous communication of problems and results, make possible unambiguous agreement on facts and their interpretations, and knit together the efforts of many independent scientists, even those who are widely separated in subject, time, and place. The renowned physicist Albert Einstein (1879-1955) noted that "the only justification for concepts is that they serve to represent the complex of our experiences." Let us take each characteristic in turn and amplify its meaning.

Some physical concepts are neither intuitive nor unambiguously understandable. In such cases, scientists believe that they can still work with such concepts by being able to operationally define them. The first characteristic mentioned in the definition is then the operational meaning of a concept. By an operational definition we mean that one can prescribe an operational procedure, that is, develop a prescribed series of actions, whereby the concept or physical quantity can be measured.

Ideally, each concept used in physical science can be clarified by some such operational definition, and that perhaps is the most important mechanism by which mutual understanding among scientists is possible; for clearly it is more difficult to misinterpret rigorously performed actions, than it is words.

Several points need to be considered regarding operational definitions. The first is that to most people everyday notions seem clear and scientific terms mysterious. Such a belief is more a product of our greater familiarity with everyday notions than with scientific terms. A little thought, however, should show that the opposite is actually true. The words of daily life are usually so flexible in definition, so open to emotional coloring, that they lend themselves to a variety of interpretations and contextual meanings. For science, this will not do. Nonscientists generally find it difficult to get used to a highly specific vocabulary and to the apparently picayune insistence by scientists on its rigorous use. However with out a rigorously defined set of terms, meaningful communication in science is impossible.

Second, operational definitions must seem to be just human convention and they do not necessarily tell us what concepts like "length," or "force," really are. Perfectly true, but therein lies their great utility. For we may make laboratory measurements suffice for discussion purposes until such time that questions concerning the deeper significance or reality of a concept can profitably be addressed.

Third, errors associated with measurements in an operational definition do not have the connotation to scientists of wrong, mistaken, or sinful acts, which exist in everyday usage of the word. Errors of measurement are merely departures from exactness. But, in time, new technology may reduce these departures so that scientists can at least strive for greater exactness in their measurements.

Finally, all physical concepts are not of equal importance. Galileo Galilei (1564-1642) pointed out that some concepts are more directly observable, and therefore can play a primary role in science. Other concepts which do not readily lend themselves to being defined operationally must play secondary roles.

The second characteristic of a meaningful concept is that most are quantitative in nature. The demand for quantitative concepts rests on an article of faith that is as ancient as it is revered. This article of faith is the belief that nature works according to mathematical laws and that observations are explained when we find the mathematical law relating observable quantities. Quoting Galileo:

"Philosophy [i.e., science] is written in that great book which ever lies before our eyes - I mean the Universe - but we cannot understand it if we do not learn the language and grasp the symbols in which it is written. This book is written in the mathematical language, and the symbols are triangles, circles, and other geometrical figures [i.e., mathematical schemes] without whose help it is impossible to comprehend a single word of it, without which one wanders in vain through a dark labyrinth."

To Galileo and his successors, mathematics provides the technique par excellence, for comprehending and seeing the unity that exists in nature first perceived by the Greeks through geometry. On a practical level, understanding and manipulating concepts is made possible by the fact that mathematically formulated ideas can be expressed symbolically in equations. Thus mathematics is both a tool for science and a universal language through which scientists communicate.

Because of the quantitative nature of concepts, the important thing for scientists about various experimental apparatus, such as weights, springs, or meter sticks, is no longer their compositions, their colors, or their histories outside of the laboratory. Since all that really matters to scientists are those mathematical relationships involving experimental objects, then the objects cease to be individual entities, and they become mathematically ideal bodies in the minds of the investigator. Their real weight is totally forgotten, and in its place we envision "a point mass." In this sense, science does not deal directly with "real bodies" but with abstractions that exist and move in a hypothetical space of precise mathematical properties. In this mathematical world of the mind, scientists can manipulate the experiment at will, eliminating in their minds all air resistance, regarding the inclined plane as perfectly smooth, or changing one aspect and leaving all other factors untouched. As seen from the outside, this idealized world, rendered in the esoteric language of mathematics and filled with simplifications and exaggerations, is quite analogous to that of the modern painter, poet, or composer. To further our understanding, let us define those concepts which are mathematical ones as those concepts taken from a mathematicized world of ideal and precise objects, and it is a realm in which actions exist as calculations.

Such concepts are often justifiable by empirical evidence, but some can not be and they find justification as coherent parts of a fruitful theory. Mathematical concepts are taken seriously only insofar as they yield new understanding of the world of our experiences.

To many scholars mathematics has moved beyond being a tool and a language in science. Mathematical equations have become a way of knowing the world that is unavailable to us in any other way. Equations possess a reality that defies language and visual images. And, they can be just as real, if not more so, in the mind of the scientist as sensual experience.

Especially useful concepts recur in a great many descriptions and laws, often in areas far removed from the context in which they were initially formulated. As you progress through the book, keep in mind our comments on the characteristics of physical concepts; for they will help you to appreciate the outlook astronomers bring to their effort in trying to understand the cosmos.

2.2. RENAISSANCE REVOLUTION IN SCIENCE

From the third century B.C. until the latter part of the seventeenth century, cosmological thought was pretty much that of the Greeks, with some mathematic refinements but no conceptual innovations. Halfway through the thirteenth century, knowledge of astronomy had spread throughout Europe as Greek manuscripts, having come by way of an earlier Arabic science, were translated into Latin in the newly founded European universities. The Renaissance blossomed in the next two centuries, ending the dominance of ecclesiastical concerns in medieval thought and beginning the development of a broader range of intellectual considerations. Renaissance scientists initiated a new era in picturing the physical world. In doing so they paved the way for important changes in scientific thought and outlook. For astronomy the most important advances were Kepler's, Galileo's, Descartes's, and Newton's concepts of motion, gravity, space, and time. After Copernicus, Kepler, and Galileo, Newton's conceptual framework introduced a whole new era in cosmological thought. Let us examine these events leading up to Newton closer.

Scientific ideas, including astronomy, proliferated after the 1450s, when the printing press was invented. Although the Ptolemaic system had been immensely successful in describing general aspects of planetary motion for over 13 centuries, the discrepancies between the observed and predicted positions of some planets had become easily recognizable by the fifteenth century. Such discrepancies prompted some thinkers to reconsider the details of Ptolemy's geocentric system. However, about the time the New World was being discovered, Nicolaus Copernicus, a Polish canon of ecclesiastical law and astronomer, questioned whether some other configuration for the planetary system might not be simpler, more reasonable, and more aesthetically pleasing than Ptolemy's geocentric one.

Around 1514 Copernicus resurrected Aristarchus's heliocentric concept and devised a new cosmology based on it. After nearly three decades of study, Copernicus's monumental book On the Revolutions of the Heavenly Orbs was published in the year of his death, 1543. Dedicating the work to Pope Paul III, Copernicus died without seeing his theory accepted. In the universities, Copernicus's book gradually became the focus of thought-provoking study, but the public reception to what was much later a revolution in the concept of the Universe was generally indifference.

Because Copernicus still believed in the Greek idea that heavenly bodies must move in perfect circles, he accounted for deviations from uniform motion by postulating a number of epicycles and other mathematical structures. His system, therefore, was not much more accurate or simpler than Ptolemy's, but the Copernican system ultimately proved to be a tremendous step in cosmological thought. The heliocentric model was as good at explaining retrograde motion and all the observed motions as was the geocentric model. As a result, in the next century this change led to acceptance of the concept that celestial physics was not a separate consideration from, but rather an extension of terrestrial physics; Isaac Newton was later to make this clear.

Appearing at an opportune time, the right man for the next advance in astronomy was Danish nobleman-astronomer Tycho Brahe (1546-1601). With financial help from King Frederick II, Brahe constructed in 1582 an observatory on the island of Hveen, about 32 km northeast of Copenhagen. There, with the most accurate pre-telescopic observing instruments ever designed, Brahe determined positions with a precision of 1 minute of arc, far surpassing any previous measurements. He observed the Sun, Moon, planets, and stars with regularity instead of haphazardly, as others had in the past. An uninterrupted record of their movements over many years was thus available for future study.

Brahe had reservations about adopting Copernicus's heliocentric theory. He accepted the idea that the five planets revolved around the Sun, but not the idea that the heavy and sluggish Earth moved. Earth's motion would be felt, he argued-and besides, a moving Earth was contrary to scriptural belief. Moreover, he was unable to detect the Earth's orbital motion by parallax in the positions of the brighter stars. Consequently, Brahe's cosmological system was a compromise: The planets orbited the Sun; the Sun and Moon, in turn, orbited a fixed Earth. There was relatively little interest in Brahe's cosmology, and it never really won a place in cosmological thought.

In the years just prior to 1600, the Renaissance and the Reformation were coming to an end. Copernicus's work was read by a few astronomers who recognized the computational advantages of the Copernican system but were not willing to take seriously its philosophical and physical implications. However, Johannes Kepler (1571-1630), the German assistant and successor to Tycho Brahe, was a devoted Copernican from his twenties on, and was destined to bring about acceptance of the heliocentric concept.

The life-long question that concerned Kepler was the nature of the clockwork that governed the celestial machinery, for he was firmly convinced that mathematical relations existed that could make sense of the planetary system. He saw the planetary system operating like a mechanical model according to its own set of mathematical laws. After years of labor, during which he rejected many ideas because they did not fit Brahe's observations, Kepler published his first two laws of planetary motion in 1609 in a book entitled New Astronomy; his third law was published in The Harmonies of the World in 1619. As Thales can be thought of as an initiator of the early period in science, Kepler can be seen in many respects to mark the beginnings of what we call modern science. Kepler developed his empirical laws from Brahe's data on Mars: "By the study of the orbit of Mars," he said, "we must either arrive at the secrets of astronomy or forever remain in ignorance of them." However, in what proved to be a revolutionary step, Kepler then generalized saying that his laws applied to all the planets, including the Earth, without ever actually verifying that this was indeed true. The expectation that the mathematical laws of science are universal in character is so readily accepted today that it is difficult to imagine just how important to science Kepler's actions were.

Kepler's work put to rest any notion that planets move in perfectly circular orbits because nature has decreed that the heavenly bodies must show perfection in their movements. Although Kepler never knew why planets move by these empirical relationships, he diligently sought a cause of which his three laws were the effect. As he stated, "I am much occupied with the investigation of physical causes. My aim in this is to show that the celestial machine is to be likened not to a divine organism, but rather a clockwork." Kepler vaguely sensed that bodies have a natural "magnetic" affinity for each other and guessed that the Sun has an attractive force. However, it remained for Newton, half a century later, to formulate a unified theory of motion that invoked gravity as the cause of planetary motion.

Galileo Galilei (1564-1642) was a contemporary of Johannes Kepler. However, for some reason, he does not appear to have been significantly influenced in his work by Kepler or Kepler's three laws of planetary motion. Galileo's approach to understanding motion in the cosmic world was through the study of terrestrial motion, especially falling bodies.

The dominant concepts of motion during the Renaissance were still those of Aristotle, who had defined motion as either natural or forced. A rock falling toward the ground was an instance of "natural motion," or the tendency of earthly materials to return to their natural place-the Earth. No cause was needed to assist such motion; it was a natural tendency. However, a thrown rock required a cause both to set it in motion and to continue it in motion. This was an unnatural tendency-referred to by Aristotle as "forced motion."

Galileo did not formulate the principle of gravity that we recognize today; Newton did that later. However, Galileo did conceptualize a force as something that brings about a change in the motion of bodies, and he saw the Earth as exerting an attractive force (that is, gravity) that influences falling bodies. He also recognized the tendency of bodies either at rest or in motion to resist a change in their motion. Thus to Galileo and his scientific contemporary in France, Rene Descartes (1596-1650), rest and uniform motion were a natural state of affairs. To change such a state, it was necessary to have a force act on the body regardless of whether it was falling straight down or moving across the surface of the Earth. A departure from uniformity in motion is now referred to as accelerated motion.

Although mechanics was possibly his most significant accomplishment, Galileo also revolutionized astronomy in 1609 by designing a telescope. As the first telescopic explorer of the heavens, though not the builder of the first telescope, he established his place in history through such discoveries as Jupiter's four large satellites, craters and mountains on the Moon, the phases of Venus, and individual stars in the Milky Way. Kepler had utilized Copernicus's heliocentric system as the basis for a dramatic new understanding of planetary motion, and Galileo gave the Copernican theory observational support. For example, he observed that on a smaller scale Jupiter's satellites moving around the planet were analogous to the planets orbiting the Sun. Here obviously were heavenly bodies not in orbit about the Earth, and here also was evidence disputing Aristotle's contention that a moving Earth would leave the Moon behind. Since Jupiter retains its satellites, then logically the Earth can move around the Sun without losing its satellite.

Theological hostility loomed over Galileo for supporting Copernican cosmology. Pope Paul V instructed his emissary, Cardinal Bellarmine, to warn Galileo against teaching or upholding Copernican doctrine, and from the Holy Office in February 1616 came the following internal memo made public later at Galileo's trial in 1633:

"The following propositions are to be censured: (1) that the Sun is at the center of the world and the Universe... Unanimously, this proposition has been declared stupid and absurd as a philosophy, and formally heretical because it contradicts in express manner sentences in the Holy

Scripture... (2) that the Earth is not the center of the world and motionless, but changes its place entirely according to its diurnal movement. Unanimously, this proposition is declared false as a philosophy."

When the more liberal Pope Urban VIII took office, however, Galileo obtained permission to discuss both the Ptolemaic and Copernican systems; he was, though, to present the latter as an unproved alternative. Encouraged by this opportunity, Galileo began work on a masterly astronomical commentary that passed censorship and was published in 1632 as The Dialogues of Galileo Galilei on the Two Principal Systems of the World: The Ptolemaic and Copernican. Powerful enemies soon convinced the Pope that Galileo had cast the Ptolemaic system in an unfavorable light. As a result, the book was officially banned, and in the year 1633 the great scientist was publicly humiliated before a papal tribunal in which he recanted his Copernican views.

Galileo spent the last nine years before his death in his villa in Arcetri, some distance from Florence, under strict house arrest. He was forbidden to publish or discuss the forbidden philosophy, although he did finish Two New Sciences and have it published in Leiden in the Netherlands in 1638. By then he was 74 and totally blind- about which he writes:

"...this Universe which by my remarkable observations and clear demonstrations I have enlarged a hundred, nay a thousand fold beyond the limits universally accepted by the learned men of all previous ages, are now shriveled up for me into such a narrow compass as is filled by my bodily sensations."

The silencing of Galileo acted to silence Catholic scientists in the south of Europe, and from there, consequently, the scientific revolution moved to northern Europe. Galileo died in the same year, 1642, which Isaac Newton was born in England.

2.3. THE LAWS OF SCIENCE

If, as Galileo believed, nature works according to mathematical laws and observations of nature are explained when we find the mathematical law relating observable quantities, we must be more specific in what we mean by a scientific law and to distinguish it from a physical concept? Let us depart once again from our historical narrative to consider the laws of science.

In 1687, Isaac Newton (1642-1727) published a treatise entitled The Mathematical Principles of Natural Philosophy, commonly known as the Principia. This monument in intellectual thought contains a remarkable passage on the rules of reasoning. There are four rules, which collectively reflect his profound faith in the unity of nature, and they were intended by Newton to guide scientists in the scientific process.

The first rule is called the principle of parsimony, and it says that scientists should make no more assumptions or assume no more causes than are absolutely necessary to explain their observations. The principle of parsimony is also known as Occam's razor, after William of Occam (1288-1348), who stated his principle of economy of thought in the phrase, "a plurality must not be asserted without necessity." The second rule is the principle of cause and effect, or the belief that what occurs in nature is the result of cause-and-effect relationships, and where similar effects are seen then the same cause must be operating. The third rule is the principle of universal qualities or the belief that those qualities, such as mass or length, that describe bodies exposed to our immediate experience also describe bodies removed from our immediate experience, such as stars and galaxies. The final rule is the principle of induction. Induction is the process of deriving conclusions about a class of objects by examining a few of them-reasoning from the particular to the more general. Deduction is the process of reasoning from the general to the more specific. The rule states that concepts, hypotheses, laws, and theories arrived at by induction should be assumed as universal both in time and place until new evidence proves the contrary to be true, as Kepler did in developing his laws of planetary motion.

These rules for reasoning are fundamental to the process of discovery of natural or scientific laws. To be more concrete in what we mean by a scientific law, let us adopt the following definition. Scientific laws are rules, preferably mathematical rules, by which humans believe nature operates, and such laws can be classified as being either empirical, definitional, or derived laws.

In their observations and experiments, scientists often synthesize their observations of phenomena by developing empirical laws, a general statement which identifies a regularity in many observations without offering a theoretical explanation for it. One good example of empirical laws is Kepler's laws.

Definitional laws are a second level of physical law, so named because these laws usually involve the definition of fundamentally important concepts. Examples of such laws are Newton's second law and the law of conservation of energy to be discussed later.

Finally, there are the derived laws which are derived from some underlying theory, such as Newton's law of universal gravitation, which is derived from Kepler's laws, Newton's three laws of motion, and the concept of "action-at-a-distance."

The scientific laws of nature are usually thought of as inexorable and inescapable, in part because the word "law" suggests an erroneous analogy with divine law. Scientific laws, being built on concepts, hypotheses, and experiments, are only as trustworthy as those concepts are complete and as those experiments are accurate. Since humans formulate scientific laws, they are neither eternally true nor unchangeable. In fact with the advance of knowledge and experience, many laws of science prove, sooner or later, to be too limited or too inaccurate. An example is the law of conservation of mass, which today we recognize as having only limited applicability.

Since its origin in Greek thought, the larger goal of science has been to explain the intricacies of nature as rationally and coherently as possible. Such an explanation does not necessarily mean attributing a motivating agent, such as God, to events, but it does mean discovering, if possible, mathematical laws between observable quantities. But how do scientists find explanation by discovering mathematical laws? Such laws may aid utilization, control, and direction, but how is anything explained thereby?

For human beings the only tools for understanding physical phenomena are the pictures, allusions, and analogies involving the primitive mechanical events of everyday life that dwell in our imaginations. Over the course of history as physical science has moved toward problems more removed from the realm of common experience, it has been necessary to supplement those mental tools with which we grasp and comprehend phenomena with mathematical concepts and laws.

Holton suggests that, "'to explain' means to reduce to the familiar, to establish a relationship between what is to be explained and the (correctly or incorrectly) unquestioned preconceptions." Modern scientists, like their ancient and medieval counterparts, do bring preconceptions to the scientific process; for example, just what we have been discussing, that nature works according to simple models or mathematical schemes. To modern science, scientific laws are an explanation of nature in that they allow scientists to incorporate mathematically the unfamiliar into the body of familiar experience. Experience shows that it requires training and repeated personal success in solving physical problems to be satisfied with and to believe that a mathematical answer explains nature.

2.4. KEPLER'S LAWS OF PLANETARY MOTION

Kepler's First Law of Planetary Motion can be stated as follows. Each planet moves in an elliptical orbit around the Sun, with the Sun occupying one of the two foci of the ellipse.

The ellipse is part of a "family" of mathematical curves known since the second century B.C. It is important in understanding the orbit of one body about another, not just planets. Roughly speaking, an ellipse is a circle with the opposite ends of a diameter pulled outward, which distorts the circle into an oval-shaped figure. The long axis of the ellipse is known as the major axis, with half the major axis being called the semi-major axis, and perpendicular to it through the center of the figure is the minor axis. The size of an elliptic orbit is set by the length of the semi-major axis. There are two points on the major axis, called the foci, whose singular form is focus, about which the figure is roughly symmetrical. In a planet's orbit, the Sun occupies one focus; the other one is empty. Since the sum of the distances from each of the foci to every point on an ellipse is a constant, this suggests a means of drawing an ellipse: Merely loop a piece of string around two tacks, which act as the foci, and wield a pencil within the loop keeping the string tight.

The farther the foci are from each other, the more elongated the ellipse; the closer together they are, the more nearly circular the ellipse. Thus the ratio of the distance of a focus from the center of the ellipse to the length of the semi-major axis, known as the eccentricity, determines the shape of an elliptical orbit. When the ratio is zero, the foci and center coincide, and the ellipse degenerates into a circle. The more elongated an elliptical orbit is, the nearer the eccentricity is to 1. The eccentricities of the planets' orbits vary from near zero for Venus to around 0.2 for Mercury. Thus planetary orbits are not very elongated but very nearly circular, which is why it was not obvious to Kepler or his predecessors that the orbits of the planets are not Plato's perfect circles.

If the planets were orbiting the Earth in circular orbits, they would traverse equal angles on the celestial sphere in equal intervals of time anywhere along their orbits. In fact as pointed out earlier, they do not actually do this, but rather traverse variable angles in equal intervals of time. In the elliptical orbits determined by Kepler, the planets' distances from the Sun, which occupies one focus, vary with their position in the orbit. Therefore, the speeds in the orbits vary from one position to another such that planets move fastest when closest to the Sun and slowest when farthest away. This variation in speed is a consequence of Kepler's Second Law of planetary motion, which may be stated as follows. The imaginary line connecting any planet to the Sun sweeps over equal areas of the ellipse in equal intervals of time.

The mean distance of a planet from the Sun is the average of the distance between the point of closest approach, called perihelion, located at one end of the major axis, and the most distant point of the orbit called aphelion, located at the other end of the major axis. The average is one-half the length of the major axis, or the semi-major axis. Therefore, according to Kepler's Second Law, a planet passes through these positions in equal intervals of time.

Kepler's Third Law is important in that it provided a means of determining the relative size of the Solar System in units of the mean Earth-Sun distance, the astronomical unit (AU). Second, it gave the mathematical relationship between orbit size and sidereal period, which was important in suggesting to Newton that gravity varied as the inverse-square of distance, as he would later demonstrate. Kepler's Third Law may be stated as follows. The square of any planet's orbital period, that is its sidereal period, is proportional to the cube of its mean distance from the Sun, that is, the length of the semi-major axis.

The orbital period is the planet's sidereal

period; that is, the time to move through 360° revolution about the Sun. The sidereal period cannot be measured directly, but as pointed out previously, it could be calculated once the synodic period was measured. Kepler believed that the constant of proportionality necessary to turn his law into an equality, was indeed constant and did not depend on the planet in any way. If this were so, when the sidereal period is expressed in units of Earth years, Kepler's Third Law would allow the computation of the mean planet-Sun distance, that is, its semi-major axis, in astronomical units. Newton later showed that Kepler's assumption about the constant was incorrect.

2.5. CONCEPTS OF MOTION

Early in this chapter, we discussed the quantitative nature of concepts in which Galileo's work played a significant role. Galileo did much to clarify quantitative concepts of motion, but most importantly he united physics and mathematics in the pursuit of an understanding of nature. It was in Galileo's consideration of freely falling bodies and accelerated motion that he revealed to physical science a new attitude toward experimentation. It was the mental ability to sweep aside those impediments in motion, such as air resistance, revealing an idealized world of motion that underlies quantitative experiences in the laboratory or the everyday world. Before we can discuss Newton's unification of the concepts of motion through his laws, we should consider these concepts individually.

A force is a push or a pull that causes a body to change its state of motion. A state of motion is the concept of a quantitative description of motion. We should note that rest, or no apparent motion, is only one possibility among many for a state of motion and should not be thought of as special. Most familiar to us are mechanical forces; these are exerted by one body in contact with another, such as a bat striking a baseball. Fields of force make up the other class of forces, and they are our response to a belief in the concept of action-at-a-distance. That is, we envision that bodies exert forces on one another without any physical contact between the bodies. Such a concept was difficult for renaissance scientists to accept, and it continues to strike us as not intuitively obvious. Action-at-a-distance is an example where defining a concept operationally proves the value of operational definitions, as we shall show in Newton's Second Law.

Science recognizes four major types of forces: strong and weak nuclear forces, which act on the subatomic scale and are responsible for structure on the nuclear scale; electromagnetic forces, which are either attractive or repulsive and give form to the world of our immediate existence; and finally gravitational forces, which are attractive, but never repulsive and are responsible for structure on the astronomical scale. Empty space between matter is no barrier to fields of force, as experience shows in the case of gravitational forces.

Every material body possesses a property called inertia, the resistance it offers to a change in its state of motion. The more matter a body has, the greater its inertia. Mass measures the amount of matter a body contains; therefore, mass is a measure of inertia. Massive bodies are more resistant to a change in their state of motion than less massive ones, as we know from common experience. Newton's attempts in the Principia to define mass were less than precise. However as we noted earlier, because of the operational nature of concepts, those mathematical concepts in a new theory that are not easily defined in terms of more fundamental concepts are still quite usable.

A material body has the same mass regardless of where in the Universe it is located, but its weight depends on its position relative to various attracting masses. Therefore weight is a measure of the gravitational force that an attracting object exerts on a body. For example, a person weighing 90 kilograms (kg) on the Earth's surface would weigh 15 kg on the Moon's surface because the Moon's gravitational pull is one-sixth that of Earth. However, the individual's mass is the same on both the Moon and Earth.

An important concept related to mass is that of compactness or density. The mass density, or just density, of matter is defined as its mass per unit volume. For example, water has a density of 1 gram per cubic centimeter (g/cm^3), while that of lead is 11.3 g/cm^3. Bodies may have the same mass but quite different densities, such as those of a feather pillow and a book. If matter is distributed unevenly throughout a body, as it is within the Earth, then the mass divided by the volume yields a mean density.

Distance is familiar in everyday experience as a measure of the space that separates material objects. However, to develop a useful description of motion, we must specify the origin from which and direction in which distance is being measured. The place where we observe and measure motion is called a frame of reference; it possesses an origin from which distances can be measured relative to some reference direction. As an example, let's consider the distances to the planets: To measure these, we must first select a reference frame, such as that with the Sun as the origin, and as our reference direction, we might choose the direction toward a given star lying along the ecliptic. This is only one of several possible reference frames that could be used to measure the distances to the planets.

Time is equally familiar from our everyday experience. Our intuitive concept of time is based on changing patterns and events in our lives-which seems different from our intuitive concept of distance, the separation between tables and chairs, for example. Time, however, like distance, can be measured from an arbitrarily chosen origin, such as a historical event. Although we can move forward and backward over a distance in space, in human experience we can only move forward in time, never backward.

For a moving body, the distance traversed divided by elapsed time is the speed of the body. If we take account of the direction of motion as well as the speed, we define the velocity of the body. A change in the velocity is known as acceleration. The change can be either in the speed or the direction of motion or both. Thus acceleration is measured as the rate of increase or decrease of a body's speed or as the rate at which its direction of motion changes. Since distance depends on a frame of reference, so also will velocity and acceleration.

An airplane's velocity, for example, is usually measured relative to the surface of the Earth, say, 600 miles per hour. We could measure it relative to the Sun, in which case the Earth's rotational and orbital velocities would have to be added to that of the airplane. Any fixed point on the surface of the Earth is continuously changing velocity, that is, accelerating, because of its rotation and revolution with the Earth. What we really want to specify in the orbital motion of planets is the velocity at one moment of time, since an instant later the velocity would be different owing to acceleration. This concept of velocity is called instantaneous velocity, and when we use the word velocity in this book, we will mean instantaneous velocity.

Simple observation shows us that there is more to defining the motion of a body than finding its velocity. For example, in the collision of two billiard balls moving with different velocities, they may simply exchange velocities so that the total quantity of motion is conserved and just redistributed between them. However, in the case of two balls of different masses moving with the same speed, the more massive one is able to transfer a greater quantity of motion in a collision than the ball with a small mass. Thus the concept of quantity of motion, or momentum, depends on both the velocity and the mass of the body. To find momentum, we multiply a body's mass by its velocity. Because momentum depends explicitly on the concepts of velocity and mass, and implicitly on the reference frame, it represents the precise mathematical definition of motion that we desire in the concept of a state of motion. Therefore, one can use these terms interchangeably.

It is not difficult to visualize philosophically the possibility that the total quantity of motion, or momentum, in the Universe is a constant. This can be stated in the form of a natural law. The total momentum of the Universe is conserved, i.e., momentum can neither be created nor destroyed, even though the interaction of bodies with each other continuously redistributes the total momentum among the individual bodies of the Universe.

This concept of conservation of

momentum was first suggested by the French philosopher and physicist Rene Descartes (1596-1650) in his book Principles of Philosophy, published in 1644. Remembering our earlier comments about the importance of mathematical laws, it becomes clear why laws of constancy, such as the conservation of momentum, are so highly prized in science. Such laws combine the most successful features of science with its most persistent preoccupation-the mathematical formulation of concepts that aid in the discovery of unchanging patterns in the chaos of experience.

In addition to a quantity of motion involved in straight line motion (translational motion), our experiences tell us that there is obviously a quantity of motion involved in rotation also. This observation gives rise to the concept of angular momentum, which is calculated as the product of the translational momentum and the distance from the axis of rotation. Reflection on your part should suggest that just as translational momentum can be conserved in the interaction of moving bodies, such as in the collision of billiard balls, so can rotational momentum be conserved. A simple consequence of the conservation of angular momentum is the observation that the spinning ice skater spins faster or slower when she draws her arms in or extends them out, respectively. The quantity of all motion in the Universe in the form of rotational motion is comparable to that in translational motion. For example, the planets rotate, stars rotate, galaxies rotate, etc. In fact, the orbital motion of the planets can be thought of as rotation about an external axis, so that orbiting planets, orbiting stars, and orbiting galaxies contribute to the rotational momentum of the Universe.

Chapter 3
Newton's Laws and Planetary Motion

In the seventeenth century, Isaac Newton united various concepts and mathematical laws of motion in one unified theory with gravity as the cause of planetary motion. Consequently, he also unified terrestrial and cosmic motion as parts of a universal motion. For Newton, gravity was the spring that ran the cosmic clockwork described by his laws of motion. The concept of gravity as underlying cause produced abundant solutions to old problems. With this one theory, Newton in his Principia accounted for the rise and fall of tides, the motions of the Moon and planets, the precession of the equinoxes, and the Earth's flattening at its poles. Moreover, it opened the door for the later discovery of the planet Neptune. No more extensive or complete system of the world has been produced than appears in Book III in Newton's Principia. The clockwork Universe persisted as a philosophical concept until

Albert Einstein, in the early part of the twentieth century, led science to see that motion does not occur in Newton's framework of absolute space and time but is relative and cannot be divorced from us as observers. Before discussing Newton's theory, let us depart momentarily to consider the nature and role of theories in science.

3.1. SCIENTIFIC THEORIES

Science is the study of the Universe that our species uses to conceive conceptual schemes or theories that are rational representations of our collective human experience. The function of theory is to help us grasp the whole picture. Following Holton, we can define a scientific theory as a conceptual scheme which we invent, or postulate, in order to explain to ourselves, and to others, observed phenomena and the relationships between them. Thus the theory brings together in one structure the observations, concepts, hypotheses, principles, and laws from often very widely different fields.

Theories are of value to science because

they unify, and in so doing they simplify. A theory is assumed to have universal applicability when it agrees with what we already know and can be repeatedly validated by future experiences to which it is applied.

Where do theories come from? Is a theory entirely suggested by a scientist's examination of data, or is it purely a product of imagination in which the scientist's mind has been set into action by contact with data? Alternatively, could the theory be the result of a combination of observation and imagination? To such questions there are no right or wrong answers. In fact, there is not even a consensus among those scholars who care about such issues.

Of great concern now as in the past to scientists and nonscientists alike is whether or not science captures reality? This long-standing question can be said to be roughly equivalent to asking which of the philosophies, realism or idealism, characterizes what scientific theories reveal. Realism is the belief that experiences that come by way of our senses must reveal a "real" world that exists independent of any human perceiver or acts of perception. The other doctrine, idealism, is the belief that there is no objective or absolute reality apart from the products of our imagination or mental constructs. According to idealism humanity can fashion only imperfect copies of that ultimate reality of our imaginations.

Idealism and realism are philosophical orientations whose origins can be traced back to Plato (427-347 B.C.) and his pupil and successor Aristotle, respectively. Thus the philosophical orientation of scientists never has and probably never will be as rigidly prescribed as one might suppose it to be. Einstein approached this difficult problem in an illuminating way. He observed that the practicing scientist "appears as realist insofar as he seeks to describe the world independent of the act of perception, [but] as idealist insofar as he looks upon the concepts and theories as free invention of the human spirit." Thus both realism and idealism may have a place in answering the question of reality in scientific theories.

Underlying theories are notions called scientific models and they have been utilized in science for a very long time. Ancient science relied frequently on analogies with the behavior and drives of organisms, and thus their theories can be said to be based on organismic models. But from Newton up to the beginning of the present century, scientific models have been mechanistic models, and much of currently accepted physical theory is a product of mechanistic thinking. However, experience in the twentieth century has shown that conceptual schemes can not always be cast in terms of some mechanical model, and there are historical examples of progress being delayed by too strong a belief in a mechanistic model. As science history has progressed, scientific models have become more mathematical conceptual schemes of mental images than purely mechanistic models. Let us define scientific models as a mental picture, or idealization, based on physical concepts and aesthetic notions that accounts for what scientists see regarding a particular phenomenon. Such a model allows scientists to predict a future course for the phenomenon in question.

The requirements of explanation and

prediction as constraints on scientific models date only from the sixteenth and seventeenth centuries. Before then, theories of nature often had to satisfy only aesthetic or theological constraints.

Whether models are devised for something as all-encompassing as the Universe or for such limited phenomena as lunar eclipses, models are widespread in astronomy. However even though models are widely used, they must be carefully distinguished from the real world. Like metaphors in poetry, their early versions may be more figurative, that is, vague approximations to the world of our experience. Our thoughts proceed, as it were, on crutches, and so depend on these mental schemes no matter how incomplete they may be.

The role of theories in science consists in correlating many separate and possibly diverse facts into a logical, easily grasped structure of thought. Such a structure may then suggest new relations that launch human imagination along previously unsuspected lines of inquiry that ultimately connect old and new facts. Theories also stimulate the making of predictions or speculations about new situations in nature, particularly in the case of quantitative theories where the prediction can be a numerical one.

History shows that successful theories are most often based on a few simple assumptions or hypotheses. And these assumptions are reasonable plausible ones to scientists, even if they are not immediately subject to test. Thus the whole tenor of a new theory is not in conflict with contemporary ideas. The fitness of a theory is most advantageously shaped and most convincingly demonstrated in a vigorous contest of ideas in which predictions play a vital role. Predictions are in essence tests of theories, and such tests may show that a theory continues to be valid, that it is in need of modification, or that it should be discarded and a new theory adopted to replace it. Thus successful theories are those that are flexible enough to grow, and to undergo modification where necessary. But if, after a full life, a theory dies, it dies gracefully, leaving a minimum of wreckage and preferably one or more descendants.

Although we are able to discern a few criteria by which successful theories are judged, one should not presume that any theory is necessarily rejected solely because it fails to meet these criteria. The decision making process is more complicated than that. With these notions of theories and their role, let us see how Newton's theory of motion shaped the evolution of science

3.2. NEWTON'S THEORY OF MOTION

Newton and most scientists after him believed in absolute time and absolute space as unchanging qualities of the Universe. Absolute here means that we can get outside the clockwork Universe and define locations in space and events in time unequivocally. In Newton's words:

"Absolute space, in its own nature, without relation to anything external, remains always similar and immovable... Absolute, true and mathematical time, of itself, and from its own nature, flows equably without relation to anything external..."

It was not until the late 1800s that Ernst

Mach (1838-1916), an Austrian physicist-philosopher, questioned the concept of absolute time, when he declared that absolute time is "an idle metaphysical conception." During his life, few took Mach's criticism seriously. Later, Einstein, in developing his theory of relativity, would be influenced by Mach's ideas. However, in the period between Newton and Einstein, the concepts of absolute time and space were pervasive influences in the development of physical science.

In 1687, Newton published his Principia through the encouragement and financial support of the great astronomer Edmund Halley (1656-1742). In the Principia, Newton formulated three laws that describe and predict the behavior of bodies in motion. As discussed earlier, both Galileo and Descartes recognized that rest and uniform motion were natural states of motion. Newton acknowledged his doubt to Galileo, Descartes, and to other formulators of physical concepts when he said:

"If I have seen further than other men, it is because I have stood upon the shoulders of giants."

Newton's First Law, often referred to as the principle of inertia, is about uniform motion. It states that a body remains at rest or moves along a straight line with constant velocity, unless acted upon by an external force.

If a body is in motion in a straight line at constant velocity, that is, its state of motion is a uniform state of motion, it continues to move along that line without changing its speed or direction as long as no force acts on the body. Another way of saying this is that its quantity of motion or its momentum remains constant. Thus the natural state of affairs in motion is for the body to continue whatever motion it has until an external force changes that motion.

Newton's Second Law deals with non-uniform motion, that is, a change in state of motion. Newton's Second Law states that a body acted on by a force will accelerate in the direction of the applied force. The greater the force, the greater will be the acceleration.

If a body's state of motion is defined by its momentum, then, if the body's mass remains constant, a change in its momentum is equivalent to a change in its velocity, and changes of velocity are accelerations. Acceleration is proportional to the magnitude of the acting force, that is, as one increases, the other does too. Another way of saying this is that the ratio of the force to the acceleration is a constant, namely the body's mass.

If a moving body is subjected momentarily to an external force, it instantaneously accelerates in the direction of the applied force. Its velocity changes to a new value, but not necessarily in the same direction as the force. If the force, however, is applied continuously, then a continuous change of velocity, or acceleration, occurs. For example, as a planet orbits the Sun, there is a continuous change of speed and direction, or a continuous acceleration. The fact that a continuous change results, tells us that a continuously applied external force must be the cause. And we know that this force is the gravitational attraction of the Sun.

Newton's Third Law deals with the relation between forces. It states that a body subjected to a force reacts with an equal counterforce to the applied force; that is, action and reaction are equal and oppositely directed.

Two forces are involved here: action and reaction. One force never acts alone. We can see equal and opposite reactive forces every day. A bird taking off from an overhead power line produces a reaction that moves the line backward. Water forced out of a lawn sprinkler produces a backward reaction that rotates the sprinkler.

3.3. NEWTON'S UNIVERSAL LAW OF GRAVITATION

Recognition that the Earth exerts an attraction on a body was not original to Newton, for Galileo recognized that the Earth exerts a force on a falling body. However, it was Newton who recognized that the pull of the Earth can extend all the way to infinity and that such a pull is a universal property of matter, inasmuch as it is possessed by all material bodies. Thus the principal hypothesis of the law of gravitation is the attraction of all particles of matter for one another.

By analyzing Kepler's second law mathematically, Newton showed that the force acting on a planet must be one directed toward the Sun. Only one kind of force would satisfy Kepler's requirement that the Sun be at the focus of the ellipse and still be consistent with Kepler's third law relating the planets' sidereal periods to their mean distances from the Sun. The force between the planets and the Sun must then be an inverse-square force; that is, the intensity of the force must weaken as the square of the distance between planet and Sun increases. Several contemporaries of Newton had suspected this relationship, but they could not prove it. To be consistent with his third law of motion, Newton had to make the gravitational force dependent on the masses of both bodies. Finally, he assembled his results in one comprehensive statement, the law of gravitation:

Newton's Law of Gravitation states that objects in the Universe attract each other with a force that varies directly as the product of their masses and inversely as the square of their distances from each other.

Newton showed that his law of

gravitation is universal by applying it to a falling apple and the Earth, to the Moon's motion around the Earth, and to the planets revolving around the Sun. He even imagined gravitation at work beyond the Solar System, a thought that was verified later.

Newton proved that a spherical body acts as if all its mass is concentrated in a mathematical point at its center. This simplifies the mathematical treatment of such bodies: The distance between their centers is ordinarily used in calculating their mutual gravitational attractions.

Since one purpose of any theory is to explain and summarize, in that respect Newton's theory was eminently successful. But, there remained an aspect which gravely bothered Newton, his contemporaries, and successors: How to account for gravity itself? What is it that causes the attraction of one body for another? Is there not some all pervading medium, later to be called ether, which somehow transmits the pull in a mechanical fashion? The very statement of these questions reflects how firmly the mind is committed to mechanistic explanations, and how abstract, and thus dissatisfying, is the mathematical argument. At the end of Book III of the Principia, Newton inserted the following:

"But hitherto I have been unable to discover the cause of those properties of gravity from phenomena [observation and experimentation], and I feign no hypotheses... To us it is enough that gravity does really exist, and act according to the laws which we have explained, and abundantly serves to account for all the motions of the celestial bodies and of our seas."

With such words, Newton denied any obligation to account for the observed consequences of gravity by additional hypotheses, for example ether, beyond those needed to derive the laws and observations. Since Newtonian gravitational theory explains such a range of phenomena, that is enough justification for accepting it. Newton was unprepared to express gravitation in terms of something more fundamental. This is not to say that he did not feel that his theory suffered by this inability, for indeed he did. In his Opticks, Newton asks whether heat is not "conveyed through the vacuum by the vibration of a much subtler medium than air, which, after the air was drawn out [as by evacuating a vessel with a pump], remained in the vacuum." This medium he imagined as "expanded through the heavens."

What we learn from this episode is that the purpose of scientific theory is not to find final causes and ultimate explanations; rather to explain observations by observable quantities and by mathematical argument. This Newton did successfully. Newton's restrain from adopting more fundamental causes was a large step toward establishing the modern conception of what is required of physical theory.

3.4. ENERGY

The so-called fundamental quantities in the study of motion are usually taken to be distance, mass, and time, and from them other concepts can be derived, such as velocity, acceleration, force, and momentum. Each of these quantities helps us to describe motion and to recognize its cause-and-effect relationships. But these quantities still do not give us a complete understanding of motion.

Recall the arguments about colliding bodies that led us to the concept of momentum, or quantity of motion. A cannonball striking a brick wall redistributes its momentum to the bricks of the wall, and they are set into motion. But momentum does not account for the destructive capabilities of the cannonball. For example, if we double the velocity of the cannonball, we find that it has twice the impulsive force necessary to set the bricks into motion but four times the destructive capability. The concept needed to explain this lacking is that of energy. In fact, no concept in science is more important than that of energy and its conservation principle.

Historically, development of the concept of energy has been long and laborious. It took more than 150 years from the first attempts at quantitative formulation by the Dutch contemporary of Descartes and Newton, Christian Huygens (1629-1695), to the point at which appropriate terminology was established. In general, energy does not have properties like those of matter, such as size, shape, and color. Moreover unlike matter, it cannot in general be said to occupy space or show inertial properties. Nevertheless, matter is but one more manifestation of energy, so we must qualify our statements by saying "in general."

How then do we define this somewhat abstract concept? We can say that energy is a measure of the ability of a physical system to perform work when the system undergoes a change. "Change" implies that we should be able to describe the system accurately before and after in order to be able to say that it has changed.

Change alone, however, is not sufficient to define energy. As human beings, we are physical systems, but changing our feelings for other human beings, for instance, does not perform work. An example of energy is what happens when a water wheel is turned by a stream as it flows over a dam; the stream performs work on the water wheel, which rotates a grindstone, which grinds grain. The energy of the stream is a measure of the ability of the stream to perform useful work.

Although energy does not in general

have the properties of matter, it can be measured and quantified. One unit used by astronomers is the erg, the amount of energy needed to accelerate a mass of one gram at a rate of one centimeter per second squared as it moves a distance of one centimeter, that is, $1 \text{ erg} = 1 \text{ g*cm}^2/\text{s}^2$.

The most important of all physical laws is the law of conservation of energy, which states that energy can neither be created nor destroyed but only transformed from one form to another.

Motion involves mechanical energy, which has two forms: one is kinetic energy, which is the energy a body has because of its state of motion; the other form is potential energy, which is the energy a body has because of its position in a field of force. A stone at the top of a hill, for example, can be said to have energy by virtue of its position in the Earth's gravitational field. If it is pushed, the stone will roll down the hill, converting potential energy to kinetic energy. We commonly make reference to such other forms of energy as chemical and electrical energy. These forms can also be understood in terms of kinetic and potential energy; but in most of this book we shall be less specific and just say "energy." The form of energy with which we are most concerned in astronomy is radiant energy.

To the physical scientist, who believes that nature works according to simple mathematical schemes, the law of conservation of energy, which says that the mathematical sum of all energies in a system can be regarded as forever constant, is a satisfying explanation for a host of phenomena even though energy is a very abstract concept. However, to the non-scientist, the same law may be regarded as an unexplained mystery, not really an explanation of anything. What is required is a personal familiarity, training, and repeated success in the solution of scientific problems to be really satisfied with a mathematical answer. As physical science has moved further from mechanistic models as tools of the mind toward mathematical ones, doubt has grown among nonscientists that science really explains. Consequently, nonscientists have focused more intently on highly-visible technology as providing, if not an explanation of the world, at least manipulation and mastery of it. Thus confusion exists and grows over what is science and what is technology.

In the decades following the publication of Newton's Principia and with the addition of the concept of energy and its conservation law, Newtonian theory expanded profoundly and new techniques were developed for analyzing planetary and stellar motions. Astronomers were able to predict the complex interactions between bodies in the Solar System and to compare these results with observations, revealing a unity between the motions of common experience and the cosmic world.

During the last half of the eighteenth century and into the next century, astronomers gathered data on all sorts of astronomical phenomena. Foremost among the observational astronomers of the eighteenth century was William Herschel. With his skill as an observer and telescope maker, he made many remarkable discoveries, including the discovery of the planet Uranus in 1781. Herschel surveyed the Milky Way, extensively cataloging many stars that orbit each other; in this he found the first positive evidence that Newton's theory of gravity is valid far beyond the Solar System.

Through the age of the Industrial Revolution and up to the beginning of the twentieth century, Newton's theory of motion reigned supreme, dispelling any doubt in the concept of a single Universe governed by a single set of mathematical laws. Newtonian physics not only shaped the development of science, but also had its impact on social and cultural development as well.

3.5. ORBITS IN THE SOLAR SYSTEM

Using his laws of motion and gravitation, Newton showed that the orbit of a body revolving around a central force like the Sun is always one of the class of curves called conic sections. These curves are called conic sections because they are formed when a plane, such as a knife blade, passes through a cone at different angles. For the ellipse, of which a planetary orbit is one example, the cutting plane intersects opposite sides of the cone's slant edge. For a circle, the plane cuts the cone at right angles to the vertical axis. The other two conic sections are open at one end: The parabola is formed when the plane passes through the cone parallel to its slant edge, and the hyperbola is formed when the cone is intersected at an angle between that for the parabola and parallel to the vertical axis.

In a parabolic or hyperbolic orbit, the body will pass by the attractive central force only once, approaching from and receding toward infinity, never to return. All objects within the solar system move around the Sun in closed orbits, or ellipses, which for most of them, are very close to circles (i.e., small-eccentricity ellipses). An object approaching the Sun from outside the Solar System would, when attracted by the Sun, travel by it on a parabolic or hyperbolic orbit. If its motion is significantly altered during a near encounter by the gravitational attraction of a planet, such as Jupiter, it could be pulled into an elliptical orbit around the Sun, in which case we say it has been gravitationally captured.

Newton also used his laws of motion and gravitation to show that Kepler's third law was only an approximation to the actual relation. He derived a modified version of Kepler's third law in which the constant of proportionality depends on the mass of both the Sun and the planet. Since the Sun's mass is so much greater than that of any planet, it is not surprising that Kepler, working with naked-eye observations, could not recognize that fact. Newton's modification is all-important, for it allows astronomers to determine masses not only for the planets of the Solar System, but also for many stars and even galaxies.

The "celestial ruler" for measuring distances in the Solar System is Earth's mean distance from the Sun, which is the astronomical unit (AU), about 150 million km. In determining the scale of the Solar System, astronomers have employed several independent techniques, of which the most accurate is that of timing the round trip of radio signals reflected from a planet. Combining this information with the planet's distance in astronomical units, employing Kepler's third law, we obtain the absolute size of the Solar System in kilometers.

A German astronomer, Johann Bode

(1747-1826), called attention in 1772 to a numerical relationship, originally discovered by Johann Titus (1729-1796) in 1766, that seemed to predict mean distances from the Sun for the then-known planets. Although not a physical law in the sense of Newton's laws, nor even an empirical law in the sense of Kepler's laws, it was often referred to as Bode's Law. Both Uranus, discovered in 1781, and the first asteroid, Ceres, found in 1801, adhered fairly well to this rule, but it broke down later when Neptune was discovered. Despite this rule's having no particular physical basis, similar types of rules relating the separations between planets seem to be characteristic of the formation of bodies in a star's gravitational field.

Because of Sun's immense mass compared to those of the planets, some 99.86 percent of the total in the Solar System, its gravitational attraction is sufficient to continuously accelerate the bodies of the Solar System towards itself. In consequence of which, the planets and other bodies orbit the Sun. The planets are similar to each other in many aspects of their orbital characteristics. For one, they revolve around the Sun in the same direction in low-eccentricity elliptical, that is, nearly circular orbits that lie nearly in the same plane. Mercury, the innermost planet, and Pluto, the outermost planet, depart most from this regularity. Between the Terrestrial planets of Mercury, Venus, Earth, and Mars, the average spacing is much smaller than that separating the Jovian planets, which are Jupiter, Saturn, Uranus, and Neptune. The planets orbit at mean distances ranging from 40 percent of Earth's distance from the Sun to 40 times Earth's distance, with orbital periods between a quarter of a year and 165 years.

Of the 125 or so natural satellites in the Solar System, all but three belong to the Jovian planets. This is exclusive of the satellites of Pluto, five of them known so far, that are technically satellites of a dwarf planet. It seems likely that more will be discovered, and in fact, many more may eventually be found since Jupiter and Saturn could gravitationally bind a lot of small bodies. Those satellites which are reasonably near their parent planet move in low-eccentricity elliptical orbits in the plane of their planet's equator and in the same direction as their planet rotates. The outer satellites usually have more eccentric elliptical orbits, which are more highly inclined to the equatorial plane of their planet. The four outer satellites of Jupiter, the most distant satellite of Saturn, and the inner satellite of Neptune have orbits that are reversed from the direction of their planet's rotation. A possible reason for these differences is that the outer satellites were captured after the planets and their inner satellite systems were formed.

Asteroids are small, rocky bodies, most of which have been found moving in elliptical orbits between Mars and Jupiter. They travel around the Sun in the same direction as the planets. However, several asteroids orbit the Sun in the vicinity of the Earth's orbit, with some in high-eccentricity elliptic orbits.

Unlike the planets, most comets move around the Sun in highly eccentric orbits with very long periods of revolution and at all angles of inclination to the plane of the ecliptic. However, there are some comets with short periods that are regular visitors to Earth's vicinity. The classic example of a short-period comet is Comet Halley which made its last perihelion passage on February 9, 1986. Comets possess very small masses in comparison to planets, which means that it is possible for the larger planets, Jupiter in particular, to alter their orbits by its large gravitational attraction. Astronomers believe comets to be "dirty iceballs" that are a conglomerate of icy materials mixed with rocky matter, whereas most asteroids are composed simply of rocky material. The cometary composition is apparently characteristic of many bodies in the outer Solar System.

Prior to Tycho Brahe's use of observations of a bright comet from Northern and Southern Europe to show an absence of any parallax, comets had been thought to be atmospheric phenomena. Brahe's efforts clearly showed the comets are well beyond the atmosphere and were probably at distances comparable to those of the planets. By the time of Halley's work on the orbits of comets, comets were generally thought to be visitors from deep space and not members of the Solar System. Halley provided conclusive evidence that comets were indeed part of the Sun's family of Solar System objects.

Halley's first acquaintance with the bright comet, that was later to bear his name, came in 1682, when it passed through perihelion on August 24 of that year. After Newton devised a method for determining a comet's orbit, Halley set to work determining orbits for several bright comets, including the one he had seen in 1682. He was struck by how remarkably similar was its orbit to those of comets observed in 1531 and 1607. In each case, the inclination of the plane of the comets orbit to the plane of the Earth's orbit was between 17° and 18°; the perihelion distance was between 0.5 and 0.6 AU. Furthermore, the direction of the motion was retrograde, that is, opposite to the sense of the Earth's orbital motion. Checking further back in time, Halley came across comets seen in 1301, 1378, and 1456, which also had similar orbits. By 1705, Halley felt confident enough to publish in his book Astronomicae Cometiae Synopsis the prediction that these six comets were one and the same comet and that it orbited the Sun in a period of 76 years. Moreover, he predicted that the comet would return once more in or around 1758. And indeed it did appear in December of 1758, passing perihelion on March 12, 1759, but unfortunately 16 years after Halley's death. The return provided indisputable proof of Newton's theory and Halley's application of it to orbit determination. In honor of Halley's prediction, the comet was named Comet Halley.

3.6. THE PRACTICE OF SCIENCE

We close this chapter with a discussion of the practice of science. In part this is because Newton did so much to determine how scientists in succeeding generations would practice their profession collectively. The success of science in the years following Newton, however, lays not so much in Newton and his predecessors having laid down a single method of work for individual scientists, but rather in a peculiar adjustment that mediates between the public and private practice of science.

Private science is that phase in the scientific enterprise of individual achievement, a single scientist or a research team working as one. Public science, on the other hand, is the entire body of science into which individual contributions are brought. Understanding the distinction between the practice of private and public science is crucial if one is to understand science.

It is during the period of individual effort that the processes of science are most difficult to characterize. In years past, when scientists felt less constrained in expressing their innermost thoughts, an irrational, mystical, or even religious conviction was often freely acknowledged as motivating their work. For example to understand the achievements of such monuments in seventeenth century science as Descartes (1596-1650), Newton, and Leibniz (1646-1716), one must understand the importance of theology in their perspective of human existence. Progress in science has so often depended on an unflinching tenacity by its practitioners. But it is precisely because private science has been able to accommodate irrational elements that the drive to discover has not been extinguished even under the most adverse conditions. We can not but conclude that intense scientific activity must provide a unique exhilaration and deep sense of fulfillment to have maintained the dedication shown by individual scientists. Henri Poincare (1854-1912) said it best when he said:

"...intellectual beauty is sufficient unto itself, and it is for its sake, more perhaps than for the future good of humanity, that the scientist devotes himself to long and difficult labors."

How can such a degree of personal commitment by individual scientists not endanger the search for objective truth? Objectivity is, apparently, imposed during the transition from private to public science, assuming it does not already exist. For only when the private stage is over and the individual contribution is formalized for absorption into public science is it that each step and concept must be clear and meaningful to the scientific community if it is to be accepted. Also non-rational tenets of individual scientists are generally so varied, so vague, and so technically inept that they can not survive simply by the lack of a basis for general acceptance and agreement. Consequently, non-rational elements are left standing in the wings when private science is ushered on to the public stage for its consideration. For what we have is a paradox in the methods of science. In the creative process, scientists may allow themselves to work and think in undecipherable ways, like creative artists, but later they must take on the role of the public scientist and speak in terms of facts, figures, and in a logical sequence of thought.

Science depends on the correct observation and classification of facts, and yet nothing can be more deceptive than facts. As noted by Aldous Huxley (1894-1963), "Facts are ventriloquists' dummies; sitting on a wise man's knee they may be made to utter words of wisdom; elsewhere, they say nothing, or talk nonsense, or indulge in sheer diabolism." Part of the problem with facts is that it is virtually impossible to discuss them without indulging in some unspoken interpretation or hypothesis. Facts cannot be discerned by themselves without intellectual tools for handling our sense impressions. Scientists do have preconceptions or "themes" and do use them in developing facts about phenomena.

This admission seems to contradict the popular notion that the first step in science is to abandon all prejudgments. However, we contend that without preconceptions one cannot conceive new thoughts. It is thought that gives scientists' eyes to perceive the world. And in short, the pattern we perceive when we note "a fact" is organized and interpreted by a whole system of thoughts, attitudes, memories, beliefs, and learned constructs. To understand even the simplest observations, each scientist places that observation in the context of a distilled wisdom of science as an institution. As so beautifully stated by Sir Michael Foster (1836-1907):

"What we are, is in part only of our making: The greater part of ourselves has come down to us from the past. What we know and what we think is not a new fountain gushing fresh from the barren rock of the unknown at the stroke of the rod of our own intellect:

It is a stream which flows by us and through us, fed by the far off rivulets of long ago."

One can not study science and not be aware of the existence of underlying themes that weave their way throughout the whole fabric of science. The strong hold that certain themes have on scientists' minds helps to explain the stubborn faith with which some scientists cling to a theory in the face of supposedly contradicting evidence. Historically, these themes have neither evolved directly from observation nor could they be incorporated as part of the logical, mathematical structure of science. One example from physics is the theme of conservation laws, such as the conservation of momentum. The conservation theme has remained a guide, even when its language has been forced to change.

Conservation laws are not the only themes. One can argue that the 2500 year course of science has been guided by such underlying themes as transformation, inevitability, and unity. From at least Thales on, the belief that nothing springs from nothing, but all exists as transformations of some fundamental entity in a multiplicity of ways, has been reaffirmed again and again. Equal to the theme of transformation is the belief in the inevitability of scientific thinking to comprehend the physical world. The theme of inevitability operates on all levels in science from the formation of scientific laws to the course of science itself. The unity theme needs no amplification since we have alluded to its presence throughout the preceding discussion. The belief that a single theory will eventually be found that encompasses all aspects of human physical existence has been a most persuasive theme in the growth and diversification of science.

In order to understand how science grows, we need a mental picture of the relational structure within the scientific enterprise. Various relational models have been asserted, for example, pyramid models and circular models; however, the one we think that most accurately represents science is to liken science to a tapestry in which the threads are the various fields of astronomy, physics, chemistry, and mathematics. These threads do not interact with each other at just one point, but they are interwoven so that they interact at numerous points. Trying to be selective in what one accepts as valid in science is to misunderstand the interlocking structure of science. Thus any attempt to remove one thread would unravel, like an old sweater, the whole fabric of science.

In much that is currently written and said about science, it is difficult, if not impossible, to distinguish science from technology. We contend that there are more than superficial differences between the two. For technology seems to be an outgrowth of our survival instincts rather than coming from our theological-philosophical origins. To help make this distinction let us define technology as what humans do to provide for their existence and it is an outgrowth of that very basic instinct for survival.

Although it is difficult to see automobiles, electronics, and all the other technical paraphernalia of our lives as by-products of the instinct to survive, that is precisely our contention. Not to misunderstand our point, we are not denying that science and technology have interacted and nourished each other throughout human existence to the point that today they are intimately entwined. For indeed they have. But science and technology have very different goals. Science is an inquiry into the nature of the world, while technology is an effort to survive as a species in that world. The fact that one can be used in the pursuit of the other does not make them the same thing.

Historically we find that paralleling questions to produce knowledge about the world have been questions of, "how we know that we know," and, "are there other ways of knowing." The field of epistemology is that branch of philosophy dealing with the nature and origin of knowledge. Ways of knowing may seem out of place in a discussion of science, but not really. For to science, knowing that you know and how you know are as important as what you know. There exists in scholarly circles a controversy over whether or not other valid ways of knowing exist besides science. Let us clarify this point.

In the theological realm, there are those who believe that ethical and moral values can only have entered and become a part of so many world societies by revelation from the Deity. In the artistic realm, it is thought that aesthetic values arise not from reason and empiricism as does science, but through enlightenment imposed by our particular collective experiences as human beings.

Granting that science is but one way of knowing, can we distinguish science and its processes from those other ways of knowing. Yes, we think we can but possibly not to every ones' complete satisfaction. Based on our discussions of concepts, models, laws, and theories, we can define science as a human endeavor to devise a way of thinking about the world and then using that way of thinking to conceive of theories that are logical representations of our sense experiences. Such theories are the most logical and economical means of representing past, present, and future events. The business of science is to trace in physical phenomena a consistent structure with order and meaning, and in this way to interpret and to transcend our direct experiences.

Many prominent scientists have contributed to the definition of science. For

Einstein, "the object of all sciences is to coordinate our experiences and to bring them into a logical system." For the Danish physicist, Niels Bohr (1885-1962), "the task of science is both to extend the range of our experience and to reduce it to order."

If it is true as Jacob Bronowski wrote that, "there is a likeness between the creative acts of the mind in art and in science," then it is not surprising to find definitions for art and philosophy that read very much like those for science. For example, the poet T. S. Eliot (1888-1965) noted that, "it is the function of all art to give us some perception of an order in life by imposing an order upon it," and the philosopher Alfred North Whitehead (1861-1947) defined speculative philosophy as "the endeavor to frame a coherent, logical, necessary system of general ideas in terms of which every element of our experience can be interpreted." This is not to say that there are not differences between science and nonscience. Quite the contrary, there are a number of points that set the two apart. Among these is the motivation that drives scientists to understand nature and her hidden beauties, to see in nature a unity in her activities, and to predict nature's course of action, while the artistic understanding of nature is primarily a realization of self and an illumination of man's place in nature.

To close this chapter, let us summarize our discussion of science by saying that science is not the impersonal technological machine that it is often portrayed as being. Science is a very sophisticated method or process of thinking about the physical aspects of our existence in this universe.

Chapter 4
Light, Optics, and Telescopes

In this chapter, we discuss the theory of light. In his Optics published in 1637, Rene Descartes asserted that the propagation of light is accomplished by the movement from one place to another of a disturbance through some all pervasive medium. He had in mind a very mechanical model, which later proved to be artificial and awkward. Descartes strenuously rejected the notion that, "something material passes from the objects to our eyes to make us see colors and light." Newton took Descartes mechanical model in a much more literal sense, and proposed a particle theory of light. Many of Descartes' and Newton's ideas either were at the time or were later shown to be inconsistent with experimental evidence. It was Christian Huygens in his book, Treatise on Light, published in 1690, who was able to answer most of the experimental facts concerning light by asserting that light was a wave phenomenon moving at very high speeds.

By the late nineteenth century,

understanding of our complex world through Newton's mechanics seemed to reveal an amazing unity. Newtonian concepts had been pushed to the atomic realm in a kinetic theory of molecular motion, and sound was well understood to be mechanical vibrations in air. Early hopes that light, electricity, and magnetism would also be explained in mechanical terms through Newtonian concepts were, however, never realized. James Clerk Maxwell was able to unify these three fields in a beautifully successful electromagnetic theory of light. But it did not encompass Newtonian mechanics. Maxwell's theory was only the beginning of a revolution in the concepts of physical science brought on by the study of light. This revolution has shown that Newtonian concepts, in spite of their many successes, are only part of the story.

4.1. WAVES AND THE TRANSPORT OF ENERGY

Astronomers have learned most of what we know about stars and galaxies by analyzing the electromagnetic radiation coming from them. Electromagnetic radiation is a form of energy, and the light to which our eyes respond is but one part of it. Since it has no material aspects, electromagnetic radiation is energy that can move through the empty reaches of the Universe.

Experience tells us that energy must be capable of being moved from place to place, that is, energy is transported by some mechanism from one location to another. Waves are one way of transporting energy. What is a wave? It is a moving disturbance. How does this account for the transport of energy? Imagine that two people several feet apart hold the ends of a rope; when one jiggles the rope, a wave travels from one end of the rope to the other. Particles are not being conveyed from one end of the rope to the other, but what is moving is a disturbance. We know that energy is transported by the disturbance because when the disturbance arrives, the receiving hand is jiggled. That is, the wave in the rope does work on the hand, giving it kinetic energy as it is set into motion. Another example of a wave is the disturbance that propagates across the surface of a pond after a stone is dropped into the water. A wave can thus be defined as a disturbance that transports energy from one point to another.

To understand waves better, consider they are described quantitatively. The distance between successive crests or troughs is called the wavelength. The number of complete cycles of the disturbance passing a fixed point per second is called the frequency of the wave. The velocity of the wave is the distance it travels per unit of time; this is just the length of each wave, its wavelength, multiplied by the number of waves passing a fixed point per unit of time, its frequency.

The last quantity used to describe a wave is its amplitude. This is the greatest height the crests reach or the greatest depth to which the troughs fall. Let us call the amplitude of the crest positive and the amplitude of the trough as negative, so that the undisturbed position is zero. The energy transported by the wave is proportional to the square of the amplitude. For example, tripling the amplitude increases the energy carried by the wave by a factor of nine, whereas halving the amplitude decreases the energy to one-fourth the original amount.

Returning to our experience of watching waves on the surface of a pond, one notes that the wave moves out in all 360° of direction across the surface. The crests of the wave are in the form of circles, and they in turn are followed by circular troughs; thus we see alternating concentric crests and troughs. Our eyes tend to follow the moving crests, which we call wavefronts. Perpendicular to the wavefront defined by the crests, that is, in the radial directions of the circular crests, is a ray that shows the direction that that portion of the wave is moving.

One aspect of waves which is of great importance to us is the concept of superposition. If we drop two stones into a pond at different points, we notice that the wavefronts defined by the crests of the two waves penetrate each other and pass unaffected through the other wave. We can state this concept as a mathematical principle of superposition as follows:

When two or more waves move simultaneously through a particular region of space, each wave proceeds independently, as if the other were not present. The resulting amplitude of the combined waves is just the algebraic sum of the amplitudes of the individual waves.

A combined wave is called a composite wave, meaning that it is composed of two or more. We can form composite waves out of individual waves that have different wavelengths, that have different amplitudes, that have been shifted slightly relative to each other, such as crest to trough, or that are going in the same or opposite directions. Many individuals speaking in a room at one time is a good example of the importance of superposition. For if the principle of superposition was not obeyed by sound waves, the multiple sound waves would distort each other and everyone would hear only incoherent noise, not speech.

4.2. ELECTROMAGNETIC RADIATION

The concept of electromagnetic radiation being waves began in 1862 when the Scottish physicist James Clerk Maxwell

(1831-1879) showed that light is energy carried in the form of a traveling wave composed of electric and magnetic fields. The electric and magnetic fields vary in intensity and are at right angles to each other and to the direction in which the wave is propagating. The electric and magnetic fields continually interact with each other to form the electromagnetic wave. While maintaining themselves, these fields continue to propagate until the energy of the wave is converted into some other form of energy. At that point, the electromagnetic wave ceases to exist. Electromagnetic radiation in the natural world occurs over a wide range of wavelengths or, equivalently, a wide range of frequencies. The product of the wavelength and frequency gives the velocity at which the electromagnetic wave travels. The amount of energy the wave transports is proportional to the square of the wave's amplitude.

In his book, Two New Sciences, Galileo suggested that the velocity of light is finite rather than infinite, but very large compared with sound velocities. The first definite evidence that light moves at a finite velocity, however, was found by Danish astronomer Ole Roemer (1644-1710). It is now-known that the velocity of light measured in empty space is 299,792 km/s (3×10^5 km/s in round figures, or 186,282 miles/s). All scientific knowledge gained thus far indicates that this is the upper limit for the velocity at which energy can be transported in the Universe. This makes the speed of light a fundamental constant of nature, which apparently has the same value throughout the Universe.

Maxwell's proposal that light is an electromagnetic wave, as we shall see, was not the last word in attempting to infer the physical nature of light from its observed properties. Visualizing light as waves spreading out from a radiating source, however, helps us to understand many aspects of it.

If we arrange the entire sequence of wavelengths for electromagnetic radiation, such that the shortest is on the left and the longest is on the right, we produce what is called the electromagnetic spectrum. Toward the short-wavelength end is the very limited portion to which our eyes are sensitive, called the visible spectrum, or visible light, or just light. The physiological response of the eye to the various wavelengths composing the visible spectrum results in what we perceive as the color spectrum.

Short wavelengths in the visible spectrum are violet, with progressively longer wavelengths producing the response we identify as the range of hues from blue, green, yellow, and orange to red in the color spectrum. Visible light is electromagnetic radiation with wavelengths between approximately 35×10^{-6} and 70×10^{-6} cm. These wavelengths correspond to frequencies between 8.5×10^{14} and 4.3×10^{14} hertz (Hz). One hertz equals one cycle, or oscillation, of the wave per second. The lowest frequencies of visible light appear red to our eyes, the highest frequencies appear violet, and between these is the rest of the color spectrum.

All types of electromagnetic radiation display those properties associated with waves; for example, all propagate in the same way with the same speed in empty space, and all transport energy. For convenience, however, we divide the non-visible portions of the electromagnetic spectrum into regions according to wavelength, such as the ultraviolet or the infrared and so on. We label these different regions not because of any intrinsic difference in the radiation, but because we have different ways of detecting radiation depending on its wavelength. Gamma rays, X-rays, and ultraviolet radiation constitute the regions with wavelengths shorter than visible light, whereas infrared, microwave, and radio radiation constitute the regions with wavelengths longer than visible light.

Because of the wide range in the numerical value of wavelengths, some units of measurement are more convenient than others for describing a region of the electromagnetic spectrum. For the visible spectrum, angstroms are convenient. An angstrom (A) is a hundred-millionth of a centimeter ($1 A = 10^{-8}$ cm). Visible radiation lies approximately between 3500 A, the violet end of the spectrum, and 7000 A, the red end. X-rays are also measured in angstroms, but infrared wavelengths are generally expressed in microns ($1 \mu m = 10^4$

A = 10^{-4} cm). Astronomers use the hertz as the unit for measuring frequency for all electromagnetic radiation.

From his theoretical study of the emission of radiation by ideal radiators, known as blackbodies, Max Planck (1858-1947), a German physicist, concluded that they do not emit or absorb radiant energy in a continuous fashion but only discontinuously in discrete units, which later were called photons. This means that the energy transported by an electromagnetic wave is not continuously distributed over the wavefront defined by the crests; on the contrary, the energy is located at discrete points, the photons, along the wavefront. In 1905, Einstein used Planck's idea of a discrete nature for the emission of light to explain a phenomenon discovered in 1887 known as the photoelectric effect. This effect cannot be understood if light has only a wave nature. Since that time, an extensive body of experimental and theoretical evidence has been collected to verify the photon concept, that is, that light does indeed exhibit a discrete nature.

What are some of the properties of

photons? They move with the velocity of light, travel in straight lines, are electrically neutral, are massless, and the energy content in each photon is inversely proportional to its wavelength. The shorter the wavelength, the more energetic is the photon; the longer the wavelength, the less energetic is the photon.

Imagine a radiating body as emitting photons of differing discrete amounts of energy in all directions. The photons are created inside atoms of the radiating body from which they receive their energy content. While traveling through space, their energy content remains constant. When photons encounter matter, they may be absorbed by its atoms, and in so doing they lose their identity by transferring their energy to the atom. The creation and destruction of photons by atoms is a classic example of the conservation of energy.

The concept of light as being simultaneously discrete photons and continuous waves seems self-contradictory and totally contrary to experience. For when we think of discrete entities, such as marbles or pebbles, applicable concepts, such as size, precise location, etc, come to mind. But for massless photons, such concepts have no meaning. As the physicist Max Born said:

"The ultimate origin of the difficulty lies in the fact (or philosophical principle) that we are compelled to use the words of common language when we wish to describe a phenomenon, not by logical or mathematical analysis, but by a picture appealing to the imagination. Common language has grown by everyday experience and can never surpass these limits."

When we resort to laboratory experiments to resolve the contradictions, we find that laboratory experiments are designed to inquire about either light's wave nature or its corpuscular nature; no experiment will simultaneously yield the discrete and the wave properties of light. Max Born concluded:

"We can therefore say that the wave and corpuscular descriptions are only to be regarded as complementary ways of viewing one and the same objective process, a process which, only in definite limiting cases, admits of complete pictorial interpretation..."

Again we must remind ourselves that the human mind tries to picture the world in terms of mechanistic models from everyday experience and not mathematical images. And we must continually struggle not to revert to the position that unless the concept is consistent with that experience, the concept is meaningless and reveals no element of reality. Wavelength is a characterization of the wavelike properties of light, while the energy content of a photon refers to its discrete nature. We can link wavelength and energy content in the mathematical equation:

$$E_{photon} = hc \, / \, \text{wavelength}$$

This is itself a strong argument in favor of the principle of duality. That is, light can be viewed simultaneously as a wave and a photon.

4.3. WAVE PROPERTIES OF LIGHT

Light traveling through empty space moves in a straight line. In everyday experience we encounter light not in empty space but passing through various media, such as the air of the Earth's atmosphere, dust or water vapor clouds, pools of water, windows, or telescopes. Under these circumstances, the velocity of light may be slowed, and the direction of a light wave may be changed. These changes are best understood through the wave properties of light rather than its photon properties.

Several properties illustrate the wave characteristics of light. One is reflection, which occurs when light strikes the boundary between two different materials, such as glass and air. When a light ray moving in air reaches such a boundary, part of it may be reflected. The reflected ray lies in the plane formed by the incident ray and the perpendicular to the boundary. The ordinary mirror, or looking glass, is an illustration of reflection.

In addition, part of the incident ray may be transmitted through the glass rather than being reflected. The transmitted ray does not, however, continue along the same straight line; it is bent toward the perpendicular. This change in direction is called refraction. If the medium into which the ray moves is denser than that from which it comes, the angle of refraction will be less than the angle of incidence. If its density is less, then the angle of refraction will be greater. A good example of refraction can be seen by placing a spoon in a glass of water. The handle looks bent at the point where the spoon enters the water because part of the handle is in the same medium, that is, air, as you and part, is under water. Thus, light must pass through the water-air boundary where it is refracted.

Light shows another wave property, diffraction, which is the spreading out of light past the edges of an opaque body. Instead of being propagated strictly in a straight line, light, like sounds waves, bends around corners. The spread is greater for longer wavelengths. Because the wavelengths of visible light are very small, we do not normally observe diffraction in the everyday world. However, one example is to observe a distant street light through an ordinary window screen. By adjusting one's distance from the screen, one can see alternating light and dark diffraction rings surrounding the street light.

Nearly all natural light sources, such as

stars, emit electromagnetic waves composed of many wavelengths. How do waves of different wavelengths add to produce a composite wave? If waves of the same wavelength from two sources are superimposed so that their crests and troughs coincide, they are said to be in phase with each other, and their amplitudes add to produce a sum greater than the amplitudes of the individual waves; the light is said to "interfere constructively." If the crests of one set of waves fall on the troughs of the other, they are said to be out of phase with each other, and their amplitudes cancel each other; the light is said to "interfere destructively." Interference is common to all types of waves. In fact, its occurrence was strong evidence that light is a wave phenomenon. Light waves of one or many different wavelengths may interfere constructively or destructively. Such waves are called composite waves, or white light, since that is the physiological response they evoke. Stars, for example, are white-light sources, although the color of the composite light from the stars may be white, red, yellow, or blue. If we can add waves together, then we must also be able to separate a composite wave into its constituent wavelengths. Indeed we can.

The surface area illuminated by an expanding sphere of light, or a portion of it, increases as the square of the radius of the sphere, that is, as the square of the distance from the light source. Since the total amount of energy leaving, say, the Sun in all directions in space is the same at all distances, the amount of radiation passing through each unit of area of the expanding sphere of light must diminish with the square of the distance.

For example, suppose that at 1 AU from the Sun the apparent brightness of the radiation over one square kilometer of surface area is one unit. At 2 AU, each square kilometer will receive 1/4 of a unit of illumination; at 3 AU, 1/9 of a unit; at 4 AU, 1/16 of a unit; and so on. This relationship between apparent brightness and distance is known as the inverse-square law of light. Thus, the apparent brightness varies inversely as the square of the distance d from the light source; that is, apparent brightness is proportional to $1/d^2$.

This law is applied in many kinds of astronomical work, as we shall see.

If an observer is moving relative to a source of waves, such as a source of light waves, or the source is moving relative to the observer, then the observer will experience or measure a change in the wavelength of the wave. For example, this familiar effect can be heard as the rising and falling pitch, or frequency, of race car engines at the Indianapolis 500 as cars approach and then move away. This phenomenon is known as the Doppler effect, named for Christian Doppler (1803-1853), the Austrian physicist who first explained it. For electromagnetic radiation The Doppler effect can be stated as being the effect seen when electromagnetic radiation received by an observer has a shorter wavelength if the source and observer approach each other, and a longer wavelength if they recede from each other. Furthermore, the amount of change in wavelength is directly proportional to the velocity along the line between source and observer.

To help us understand, suppose a stationary light source, such as a star, is radiating concentric waves of one wavelength in all directions. Then observers in any direction, if stationary, would see successive crests of the wave passing them at the same rate at which they were emitted by the star. If, however, the star begins to move at uniform velocity observers along the line of motion would see crests passing them at rates different from that with which they were emitted. To see why, consider the water waves at the bow of a boat. In the direction of motion, the wave crests appear more compacted, and in the opposite direction they will appear more separated. Thus, for the light from stars, the wavelength is shifted toward longer wavelengths, or redshifted, as a star recedes from an observer, and toward shorter wavelengths, or blueshifted, as a star approaches an observer. Observers located at right angles to the moving star, would detect no change in the rate for crests passing them. Observers elsewhere would notice some change, the amount depending on the angle between their radial direction to the star and the line of motion.

It is immaterial whether the light source

is in motion, or the observer, or both. That is, the size of the Doppler Effect found depends only on the net relative motion along the line of sight between the source and the observer. The amount of the wavelength shift due to the Doppler Effect is directly proportional to the velocity of approach, blueshift, or recession, redshift, as long as the relative velocity is well below the velocity of light. The constant of proportionality is the ratio of the non-displaced wavelength to the velocity of light. This means that the wavelength shift is greater the longer the wavelength of the radiation. Since all bodies in the Universe are moving, the Doppler Effect is an important tool for detecting and measuring the amount of motion along the line of sight.

4.4. OPTICAL TELESCOPES

Having a foundation now in the properties of light, we can proceed to consider the means by which astronomer collect and analyze light from astronomical bodies. The basic collecting instrument is the telescope which we will discuss in this and the following sections. Telescopes, as well as other optical instruments, depend for their operation on the wave properties of light. On the other hand, the analysis of light to extract information about the light source requires some further study of matter and radiation.

In optical astronomy, astronomers work with the image of the light source formed by the principal image-forming part of the telescope, which is called the objective; the objective is either a lens or a mirror. Light rays from the light source are refracted in passing through a lens and are reflected from a mirror. The image is formed where light rays converge to a position known as the focus. The focal length of the objective is the distance behind the lens to the focus or the distance in front of the mirror to the focus. The telescopic image of a star is just a point of light, while that of an extended object, such as the Moon, is extended but inverted.

In telescopes using either mirrors or lenses, an eyepiece, another small lens, is used to magnify the image much as a magnifying glass enlarges small print. Or instead of an eyepiece, a photographic plate may be inserted into the focal plane, transforming the telescope into a camera, where the objective serves as the camera lens. The advantage of photography over observing with the eye is that time exposures can record fainter objects than those the eye sees, and in addition, the photograph is available for later study.

The image formed by either a lens or a mirror has certain properties that depend on the diameter of the objective, or aperture, and its focal length. One property is the size of the image. Since the image of a star is a point, size is not an important consideration. But for an extended object, such as a galaxy, the image size depends on the angular size of the galaxy on the sky and on the focal length of the objective.

The brightness of the image is important because it determines whether the object can be seen and how long it will take to photograph. The brightness of an image of a star depends on how much light is intercepted by the objective. Hence its brightness is proportional to the area of the objective or to the square of the aperture.

Doubling the aperture but leaving the focal length the same increases the area of the objective or its light-gathering power four times, concentrating four times as much light into the same-size image.

When photographing a galaxy, the image brightness depends on the amount of radiant energy per unit area of the image. The objective's area, or the square of the diameter of the aperture, still determines the total amount of energy collected, but the total energy is distributed over an extended image. Therefore it can be demonstrated that the larger the image's area, the smaller the energy per unit of area. The image size of a galaxy increases in proportion to the focal length, so for a given telescope aperture the surface brightness of the image decreases as the focal length is made longer.

How well a telescope can discriminate between two objects close together on the sky or can bring out fine details in an extended object is called its resolving power. Because of the wave nature of light, the image of a star is actually a diffraction pattern. That is, it appears as a bright central spot, called a diffraction disk, surrounded by progressively fainter rings. When the diffraction patterns of two stars that are close together no longer overlap, we can see separate stellar images. The larger the aperture of a telescope, the smaller the diffraction disk of each image. A large aperture therefore improves the resolution of closely adjoining features by making the diffraction effect of adjacent objects overlap less. We define resolving power as the smallest angle between two close objects whose images can just be separated by a telescope. This critical angle is directly proportional to the wavelength of the observed radiation and inversely proportional to the aperture of the objective.

4.5. REFLECTING AND REFRACTING TELESCOPES

Telescopes that use lenses for the objective are known as refracting telescopes, whereas those which employ a mirror are called reflecting telescopes. The objectives of early refracting telescopes could not form sharp images because of a condition known as spherical aberration; these single lenses also failed to bring all colors to a common focus, a condition known as chromatic aberration. These conditions can now be minimized by using a compound lens, or two lenses of different types of glass cemented together, as the objective in refracting telescopes.

Spherical aberration also occurs in reflecting telescopes. If the surface of the mirror is parabolic rather than spherical, then spherical aberration is eliminated, although some minor deficiencies still remain.

Why are the big modern telescopes of the reflecting type? Reflecting telescopes have many advantages over refractors. The reflecting telescope is free from chromatic aberration, making it ideal for all-purpose photography and spectroscopy. Also, since a lens must be supported by its edges, there is a limit to the size of a lens that will not break from its own weight. But a mirror can be supported both at its edges and from the back, and such a means of support allows much larger mirrors to be built than lenses. The largest refractor has an aperture slightly over 1 m, but the largest reflector is 8 m in diameter.

There are other advantages to reflectors. For example, the glass for the mirror in a reflecting telescope does not need to be as optically pure as that required for a large lens because the light reflects off the front surface and does not pass through the mirror, as it does through a lens. In addition, a mirror has only one surface that must be painstakingly ground and a compound lens has four surfaces. To counter changes in temperature that would affect the focal length of the reflector, large mirrors are constructed of fused quartz or of a zero-expansion pyroceramic material. The mirror's surface is coated with a thin layer of highly reflecting aluminum that is replaced many times during the life of the telescope.

Reflecting telescopes can be designed for many kinds of astronomical work through choice of the focal arrangement to suit the type of observation. For photography, photometry, and spectroscopy of faint objects, the prime focus is best because its small focal length lessens the exposure time required. The Newtonian focus, most useful for small telescopes, is now little used by professional astronomers. In both these arrangements, the observer works at a considerable distance above the observatory floor, since both focal positions are near the entrance of the telescope.

In the Cassegrain focal arrangement, a secondary mirror at the entrance of the telescope is used to slow the rate at which light rays converge after reflecting off the objective mirror, effectively increasing the telescope's focal length. The secondary mirror reflects the converging rays to the bottom of the telescope and through a hole in the objective mirror to a focus behind the objective. This is a much more convenient observing position since it is near the floor and behind the telescope. Of all the observations made with the 5-m Hale telescope on Palomar Mountain, 75 percent are from the Cassegrain focus.

One might think that putting the secondary mirror and its supports or the observer's cage for the prime focus into the path of the light rays would obscure part of the image, but the only effect is to cut down the amount of light reaching the objective. The loss is small, and the quality of the image is not affected.

Equipment that is too heavy and bulky to be attached to the back of the primary mirror or is sensitive to changes in gravity as the telescope moves can be placed in a room below the observatory floor. Through the use of an auxiliary flat mirror, the long converging beam from the primary mirror can be diverted down the hollow polar axis around which the telescope rotates and into the room below. With this Coude focal arrangement, the focal position always remains the same no matter which way the telescope points.

An optical telescope, in order to follow an object as the Earth's rotation carries it across the sky, must be free to move. In order to track stars accurately and to permit a telescope to be pointed in any direction, an equatorial mounting system is used for most telescopes. This system has two axes of rotation: The telescope can be made to rotate in an east-west sense, called hour-angle, around its polar axis, which is aligned with the Earth's axis of rotation; the declination axis, which is perpendicular to the polar axis, is used to rotate the telescope in a north-south sense.

Large telescopes are usually pointed by a computer from an operating console and guided thereafter with hand controls. Once a large telescope is properly pointed, the computer operates as a clock to slowly turn the telescope westward around its polar axis at the same rate as the Earth turns eastward, thereby keeping the area of interest always in the telescope's field of view. The great simplicity in an equatorial mounting is that tracking requires continuous motion about only one of its two axes. The disadvantage, which applies only to the largest telescopes now in operation and planned for the future, is that the polar axis is inclined in the Earth's gravitational field and must rotate on one edge of its end. In this position, gravity creates a large mechanical stress on the polar axis that presents a very difficult engineering problem.

One means of preventing some of the mechanical stress on the polar axis is to align it with gravity. Such a mounting is known as altazimuth mounting; with it a telescope rotates about a vertical axis and about a horizontal axis. This mounting's disadvantage is that unlike the equatorial mounting, it must turn continuously about both axes at the same time in order to track a star. When the telescope approaches the area of the sky directly overhead, continuous tracking becomes virtually impossible. Even with this disadvantage, the altazimuth mounting will be the primary mounting for very large telescopes to be constructed in the future.

The geographic location of an observatory is important in order to obtain all the capability built into the telescope and its mounting. An ideal site for an optical observatory is a mountaintop where turbulent motions in the atmosphere are minimal, the air is dry and transparent, and the sky is dark. The southwestern part of the United States, or similar spots along the West Coast, is a most satisfactory location. In addition such sites have many clear days and nights. Kitt Peak National Observatory is located in such a site, on a mountaintop, about 70 miles southwest of Tucson, Arizona.

The principal problems in building very large telescopes on the Earth's surface are cost and construction time. A new 5-m single reflector could cost about $50 million and take up to 5 years to build. Clearly some dramatic changes in design were needed to lower cost and construction time. One type of newer telescope design is called a multiple-mirror telescope. It uses a mosaic of independent mirrors of smaller size coordinated by laser beams to collect and focus light in order to simulate the collecting ability of a large-aperture single mirror. Such telescopes consist of a circular array, for example six identical 1.8-m mirrors on an altazimuth mounting; the array has light-gathering power equivalent to that of a 4.5-m single mirror. The six mirrors are not thick solid ones but are of a newer lightweight design. They are partially hollow, which require a smaller mechanical structure to move them. Thus the cost of such multiple mirror telescope is about one-third that of a conventionally designed telescope, and it requires less time to build.

A multiple mirror telescope is not the only newer design. There are other newer designs for future large telescopes.

Finally, it should be noted that orbiting

telescopes like the Hubble Space Telescope are not be limited by light losses produced by the atmosphere. Thus, they can operate at shorter and longer wavelengths than the visible portion of the electromagnetic spectrum. The Hubble Space Telescope can actually see into the ultraviolet regions of the electromagnetic spectrum. The future James Webb Space Telescope is being designed to see into the infrared portion of the electromagnetic spectrum.

Chapter 5
Matter and the Study of Radiation

Matter and the radiation it produces or annihilates are of vital importance to astronomy. The physical nature of various astronomical bodies dominates the efforts of astronomers, and it is through the radiation emitted by these bodies that we are able to say what they are like. To a great extent, work in astronomy today is what has been called "the practice of radiation diagnostics," that is, the collection, analysis, and interpretation of the radiation emitted all across the electromagnetic spectrum by astronomical bodies. Therefore, we want to continue in this chapter our investigation into the nature of matter, radiation, and its detection.

5.1. STRUCTURE OF MATTER

Some of the earliest philosophical speculations, such as that by Thales, were concerned with what the material world is made of. Is each substance, such as rock or wood, infinitely divisible so that its subdivisions always yield the same properties as the whole substance? Or is there some level of structure below which the subdivisions will show new properties and forms? The Greek philosophers Democritus (460-370 B.C.), Leucippus (c. 440 B.C.), and later Epicurus (341-270 B.C.) suggested that the material objects of our experience are actually made up of fundamental units. They called these units atoms, and they visualized them as indestructible, indivisible, infinite in number and variety, and capable of being assembled into various forms and shapes. However, the most important aspect was the concept that matter was never born from nothing, but sprung only from new combinations of atoms. This laid the foundation for the concept of transformability as one of the great themes in science. Along these lines, the Roman poet Lucretius (99-55 B.C.) suggests that, "All nature then, as it exists by itself, is founded on two things: there are bodies and there is void in which these bodies are placed and through which they move about...." Thus physical existence is composed of two realities; everlasting particles, atoms, and what we may loosely call vacuum, or the absence of particles.

Although, a number of early conceptual schemes were based on small indivisible particles, it was not until the period between Robert Boyle (1627-1691), British chemist-physicist, and the death of the French chemist Antoine Lavoisier (1743-1794) that the concept of the chemical element was to arise and give a new, vital, and precise meaning to atoms. The concept of the atom was not widely accepted until the English scientist John Dalton (1766-1844) developed his atomic theory of gases around 1800. Dalton stated:

"Matter, though divisible in an extreme degree, is nevertheless not infinitely divisible. That is, there must be some point beyond which we cannot go in the division of matter....I have chosen the word atom to signify these ultimate particles....[which for] all homogeneous bodies are perfectly alike in weight, figure, etc. In other words, every particle of hydrogen is like every other particle of hydrogen..."

Dalton also saw matter as composed of combinations of atoms of a finite number of chemical elements. And, he reaffirmed the concept that transformation involves the rearrangement of atoms, but neither their creation nor destruction.

Modern science has shown that even atoms can be subdivided into more fundamental units. In 1897, the English physicist J. J. Thomson (1856-1940) identified the electron, which carries a unit of negative electrical charge, as a constituent of atoms. Also in England early in this century, Ernest Rutherford (1871-1937) demonstrated that most of the atom is empty space and that nearly all its mass is concentrated in the nucleus. The principal constituent of the nucleus is the positively charged proton, which was identified by Rutherford about 1919. By 1932, a second particle having no electrical charge and called a neutron was also shown to reside in the nucleus by the British physicist James Chadwick (1891-1974).

The early 20th century picture of atoms was based upon the work of Danish physicist Niels Bohr (1885-1962) and known as the Bohr model of the atom. This has a nucleus made of protons and neutrons, with electrons in orbits surrounding the nucleus. The chemical identity of each atom of an element is determined by the number of protons in its nucleus, which in turn establishes the element's atomic number. The simplest nucleus is that of hydrogen, with one proton and atomic number 1; the atomic number of helium is 2, that of lithium is 3, and so on. Protons have a unit of positive electrical charge equal and opposite to the negative charge of electrons; the mutual attraction between positive and negative charges holds atoms together. Neutral atoms have as many electrons as protons. When there are fewer electrons than protons, the atom is known as a positive ion. Atoms bond together to form molecules, with the number of atoms varying from two, as in the oxygen molecules we breathe, to many millions, as in complex hydrocarbons that compose biological life. This conceptual model, with some additional refinements, accounts satisfactorily for the periodicities in the physical and chemical properties used to arrange the elements in what is called the periodic table of elements.

Atomic nuclei are made up of from 1 to about 260 protons and neutrons. The atom's mass is approximately that of the nucleus, since the mass of either protons or neutrons is almost 2000 times greater than that of electrons.

Although the nucleus of any particular element contains a fixed number of protons, the number of neutrons may vary from a few more to a few less than the number of protons. These nuclei with different numbers of neutrons and consequently different masses are called isotopes. All the elements in the periodic table, except for 20, possess two or more isotopes, so that their atomic weights depend on the relative abundance of each isotope in nature.

The search for the ultimate indivisible constituents of matter began before Democritus and really continues even today. Physicists in the latter portion of the previous century had their hands full discovering more then 100 different subatomic particles. In the 1960s, physicists began to incorporate all of these in a model now known as the standard model of the atom. In the standard model, protons and neutrons are themselves composed of even more basic particles, called quarks by Murray Gell-Mann (b. 1929). Is there an ultimate end to the divisibility of matter? On a conceptual level, most scientists would like to think so, and those that study string theory believe that quarks themselves are made of conceptual objects called strings. If string theory is correct, these represent the most basic unit of matter in the universe, themselves considered as one dimensional packets of energy.

In any event, ordinary matter exists in three states. In solids, atoms are bound to permanent positions relative to each other; in liquids the particle bonds are weak and temporary. By contrast, in gases there is no significant bonding between atomic particles and the particles have no permanent positions relative to each other. The particles of a gas can be molecules, which consist of two or more atoms, atoms themselves, or ions and electrons.

Most of the ordinary matter in the Universe is either in the form of gas or a gas composed of free electrons and positive ions called plasma. Each atom in a plasma can be stripped of one electron, so that there are two independent particles per atom; or stripped of two electrons, three particles per atom; and so on. Only some of the atoms in a plasma may lose an electron, or in the extreme case, all the atoms may lose all their electrons. Such a high degree of ionization is generally the result of very high temperatures.

In the realm of atoms, as in a gas, gravity does not cause changes in motion, as it does in the macroscopic world. The atomic world is dominated by electromagnetic forces. As with the force of gravity, the intensity of electric and magnetic fields weakens as the inverse square of the distance from their source. At first glance, this might suggest that Newtonian mechanics ought to describe motion in the atomic domain, gravity as a cause simply being replaced by electromagnetic forces. Such is not the case in general, however, and the mechanics of the atom is called quantum mechanics. Its details go beyond our needs in this book, so we only point out that motion in the atomic world has a discrete nature rather than the continuous characteristics of our everyday experience. This radical departure in the fundamental cause-and-effect relationship from the macroscopic realm, in which we live, to the atomic realm, is in the same conceptual vein as Einstein's concept of the photon as noted in his photoelectric effect paper published in 1905. That is, the concepts underlying quantum mechanics and the photon are the same.

In gases, atomic particles dart about rapidly, colliding millions of times each second and changing their directions of motion just as frequently. Each gas particle has a kinetic energy proportional to the product of its mass and the square of its velocity. After a collision, the velocity can be either greater or smaller than it was before the collision; the kinetic energy of each particle changes in its repeated collisions. Collectively, however, the gas particles will have some average kinetic energy, that changes only when energy is added to the gas or removed from it. Another way of saying this is that the average kinetic energy changes when the gas is heated or cooled. Thus what we call heat or thermal energy is no more than the collective kinetic energies of atomic constituents.

Temperature is a measure of the average kinetic energy of gas particles. The motion of the particles composing a body, such as ice or water or water vapor, is called random thermal motion. It increases as the temperature goes up, and it decreases as the temperature goes down. Absolute zero is reached when the average kinetic energy is zero. Seen in terms of the motion of the particles in a gas, temperature is a measure of that motion: The greater the temperature of the gas, the greater the random thermal motion.

Temperatures in astronomy are usually measured on the absolute, or Kelvin (K), temperature scale. In this system, there are 100 equal divisions or degrees, between the freezing point of water at 273 K, and the boiling point of water at 373 K.

When we heat the air in a vessel, we are increasing the kinetic energy of each particle, which means a higher average kinetic energy. As a result, the gas particles move about faster, and they collide more frequently and more violently with their surroundings. If the density of the gas particles is quite large, then the hot gas can transport a great deal of thermal energy, or heat, from one place to another. The flow of energy in nature is from regions in which the energy content is high to those in which it is low; natural physical processes tend to even out the energy content.

5.2. KIRCHHOFF'S LAWS

Our understanding of the nature of matter, particularly gases and plasmas, as to its chemical composition, temperature, density, and motions, can be enhanced by examining the radiation either emitted or absorbed by matter. In this and the remaining sections we will consider how radiation conveys information about matter, and what techniques are used by astronomers to collect and measure radiation.

Until about a century ago, astronomers were concerned primarily with the positions and motions of celestial bodies. They knew little or nothing about the physical nature of these bodies and were really not able to find out. Today, however, concern with the physical nature of celestial bodies, which is the field of astrophysics, is one of the most actively pursued areas in astronomy. This change in orientation is primarily the result of two developments in of our understanding of light.

The first development was the invention of the spectroscope in the late 1850s by chemist Robert Bunsen (1811-1899) and physicist Gustav Kirchhoff (1824-1887). Earlier in 1704, Newton, in his Opticks, had described how one saw a rainbow of colors when sunlight was passed through a prism, the principal component of a spectroscope. After Newton, both William Wollaston (1766-1828) in England and Joseph Fraunhofer (1787-1826) in Bavaria used prisms to investigate the colors emitted by various chemical elements. Thus the spectroscope is essentially a device used to separate white light into its component colors.

The second development was the recognition that under certain circumstances each chemical element emits a specific set of colors that is peculiar to it, much like a person's fingerprints. As early as the 1830s this fact was suggested in connection with the presence, identity, and abundance of different elements in ores. The real beginnings of the field of spectroscopy occurred in the last half of the nineteenth century in the laboratories of Bunsen and Kirchhoff at the University of Heidelberg.

From his experiments, Kirchhoff was able to formulate three empirical laws of spectroscopic analysis. These laws describe the physical conditions under which matter will produce light having one of three different spectra of colors. One of the first astronomical applications of these laws was in trying to determine the chemical composition of the Sun and stars.

By 1864, the English astronomer Sir William Huggins (1824-1910) had identified nine elements in the bright star Aldebaran. Four years later, in 1868, Sir Norman Lockyer (1836-1920) detected an element in the solar spectrum that was unknown on the Earth at that time. It was later found in natural gas, but it still carries its solar name, helium. In the years following, more elements were identified in stars. In addition, it was also discovered that the spectra of colors in the white light coming from stars contain sufficient similarities that the stars can be arranged into spectral classes. There are, however, small but important differences in the spectra of colors from star to star.

By the beginning of the 20th century, an important tie had been developed between the researcher in the laboratory and the astronomer in the observatory. From this collaboration came a new way of perceiving nature and this has fundamentally altered our conceptual view of the Universe. The extension of spectrum analysis to radiation in parts of the electromagnetic spectrum other than the visible and the ability to move above the Earth's obscuring atmosphere to view radiation coming from the depths of the Universe are the Rosetta stone of today's astronomy.

Just as sound waves of various wavelengths are transported simultaneously through the same region of air, electromagnetic waves of different wavelengths can move through the same point in space and superimpose to form composite waves, or white light. In turn, white light can be dispersed, or separated, into its component colors, or wavelengths, to form a spectrum. The study of spectra is called spectroscopy. Let us briefly explain how we can accomplish the dispersion of composite light by using a triangular piece of glass called a prism.

In the refraction of a ray of light, the angle through which light is refracted depends on wavelength; the angle of refraction is greater for shorter wavelengths than it is for longer ones. Consider white light passing through the slit of a narrow diaphragm and then through a glass prism. The light separates into its component wavelengths, since short-wavelength light is refracted through larger angles than is long-wavelength light. Thus waves of different wavelengths go off in different directions. The result, when imaged on a screen, is a rainbow-colored sequence of images of the slit containing an image for each wavelength present in the white light.

Another means of dispersing white light

to produce a spectrum is the diffraction grating. Unlike the ordinary glass prism, which is transparent only to visible and infrared radiation, the grating is useful over a broad spectrum, from X-ray to infrared wavelengths. In its simplest form, the diffraction grating is a plate containing a very large number of very narrow parallel slits uniformly spaced at distances that are only a few times the wavelength of light. By "large number," we mean many thousands of slits per centimeter. The spectrum is viewed in the direction of the light source. Since the amount of bending, or diffraction, of electromagnetic waves at each slit depends on wavelength, composite light is separated into its component colors.

When we analyze white light from various astronomical sources for its color composition, we do not always find a continuous rainbow-colored sequence of wavelengths. Spectra can be classified and interpreted according to laws formulated by Kirchhoff more than a century ago. The three basic types of spectra, i.e. continuous, emission, and absorption, and the physical conditions under which they are formed, are given by Kirchhoff's laws.

Kirchhoff's First Law states that the spectrum of a radiating solid, liquid, or highly pressurized gas is an uninterrupted sequence of wavelengths known as a continuous spectrum.

Kirchhoff's Second Law states that the spectrum of a radiating rarefied gas is a set of discrete, or isolated, wavelengths whose appearance is a series of bright-colored lines that form a pattern characteristic of the chemical composition of the gas, and is known as an emission or bright-line spectrum.

Kirchhoff's Third Law states that light from a radiating source producing a continuous spectrum will, if it passes through a cooler gas, have certain specific wavelengths characteristic of the cooler gas removed from the spectrum. The spectrum appears continuous except where it is crossed by dark lines, which indicate that these wavelengths have been removed, and it is known as an absorption or dark-line spectrum.

There are many common examples of light sources whose spectra are one of the three basic types. For example, the glowing filament of an incandescent electric light bulb produces a continuous spectrum. A neon sign is an example of an emission spectrum. The spectrum of a gas composed of molecules is many sets of very closely spaced spectral lines known as emission bands. And, as a final example, the spectrum of the Sun and most stars is an absorption spectrum.

We will see additional examples of each type of spectrum many times in the remaining chapters. The important point to remember is that the type of spectrum for a light source tells us something about the conditions in and around that source.

An astronomical light source, such as a star or a gaseous nebula, contains a mixture of chemical species, each either emitting or absorbing its own set of wavelengths of electromagnetic radiation. By knowing what wavelengths are emitted by different chemical elements from laboratory studies, astronomers can identify individual elements in the light source from the measured wavelengths of its spectral lines, regardless of whether they are emission or absorption lines.

Identification is done in the following way: Light from a celestial body is collected by a telescope and then passed through a spectrograph in order to disperse the white light from the source and form its spectrum. The photographic plate on which the spectrum is recorded is called a spectrogram. As a standard against which unknown wavelengths in the astronomical spectrum can be measured, an emission spectrum of a known gas, such as neon or vaporized iron or titanium, is placed above and below the astronomical spectrum. The mechanism for placing the laboratory spectrum on the astronomical spectrogram is a part of the telescope and spectrograph. With these comparison lines of known wavelength the astronomer can determine the unknown wavelengths of the astronomical object's spectral lines.

Kirchhoff's laws of spectrum analysis tell us about the general physical conditions of the light source. And if the spectrum of the light source contains absorption or emission lines, we can measure their wavelengths and identify the chemical elements that are present.

Can more detailed information about the light source be found? Suppose we want to know the temperature of the light source. Can this be done? Yes it can, for special types of light sources known as ideal radiators, or blackbodies. In Chapter 14 we shall outline the basis for obtaining more detailed information about stars, such as density and rotation.

5.3. RADIATION MEASUREMENTS

Before discussing the instruments used with optical telescopes to obtain quantitative measurements of radiation, let us consider briefly the most important component of these instruments: the radiation detector. Telescopes are capable of collecting light over a wide range of wavelengths, but it is the radiation detector that actually determines what the telescope "sees." One radiation detector with which we are all familiar is the human eye. Since the eye is so familiar to us, we can use it to illustrative those properties of radiation detectors which are of interest. Such properties are:

- the wavelength region to which the detector is sensitive;
- the variation in response of the detector over that wavelength region; and,
- the variation and range of detector response to different levels of illumination.

The eye is sensitive to the narrow-wavelength region between about 3500 and 7000 A. However, the eye does not respond equally to all colors in the visible spectrum. It is most sensitive to the middle of the wavelength region, the green region, and sensitivity drops to zero toward either the violet, shorter wavelength, or the red, longer wavelength regions.

The variation and range of detector response is the way in which the eye responds to one photon or to a flood of photons. Experience tells us that the eye does not respond in the same way to both. For the eye, a minimum number of photons is required, depending on their wavelength, to make it respond. In other words, there is a limit to how faint a light source we can see, and that visibility limit depends on whether we are looking at violet, green, or red light.

All of us have experienced the loss of response when the eye is exposed to a very bright light. In such cases, the eye saturates. That is, it no longer responds, and no scene is visible, just an intense and painful brilliance. To be useful, the range between minimum and saturation of visibility should be quite large, say, a factor of 100 or 1000. If anywhere between the lower and upper limits of the eye's response, we double the number of photons from a light source; do we observe that the light is twice as bright? The answer in general is no. By and large, over the eye's range of response, doubling the stimulus does not double the response; in other words, we say that the response is nonlinear. This concept of linearity is important because in seeking the amount of radiant energy emitted by an astronomical source, astronomers usually compare the unknown light source to one of known energy output. Thus they have to know how their radiation detector responds to increasing or decreasing numbers of photons. Let us now switch and consider two other radiation detectors, the photographic emulsion and the photoelectric device.

Photographic emulsions record photons by undergoing a photochemical change that will ultimately deposit silver on a glass plate or acetate film. Various photographic emulsions can be manufactured such that they will respond to different wavelength regions within and beyond either end of the visible spectrum, which makes the photographic emulsion more versatile than the eye. Photographic emulsions, like the eye, are nonlinear in their response; they have a rather complicated response depending on position in their response range.

Photographic emulsions possess a significant advantage over the eye in that they build up the image by storing their response over time. Thus time exposures allow the astronomer to collect information on a photographic plate about very faint light sources that cannot be seen with the eye through the same telescope. How faint a star can we photograph? The telescope's aperture sets the initial limit. Ultimately, however, the limit is set by a weak illumination coming from the whole night sky. This background radiation comes from two sources: starlight scattered by the Earth's atmosphere and the airglow emitted by the atmosphere itself. The inherent disadvantage of photographic emulsions is that its photon-capturing efficiency is low. It can record only one or two percent of the incident photons, those which activate the light sensitive coating. Facing such inefficiencies, astronomers have found other types of radiation detectors to improve the performance of telescopes.

The photoelectric device is an application of the photoelectric effect. The basic principle is to use photons to eject electrons from a metal surface by exposing it to a beam of light and then to measure the number of electrons liberated with electronic circuitry. A photoelectric device, like photographic emulsions, can be made to respond to different wavelength regions by using different metals for the surface of a device. The biggest advantage of the photoelectric device is that it has a very large range in its response. In addition, its response is linear to the number of incident photons. With modern electronics it is possible to adapt the photoelectric device to count individual photons or to use a mosaic of devices to form a picture much as a photographic plate does.

The photographic plate and the photoelectric device enhance our ability to detect light from astronomical sources of different brightnesses, but they are not basically analyzing instruments. To analyze light, astronomers must equip an analyzing instrument with either of these detectors and attach it to a telescope. The two basic types of analyzing instrument are the spectrograph and the photometer.

The spectrograph disperses composite light from the source into its component wavelengths so that we can, for example, determine the elements that compose the light source. Spectroscopy, which is the study of the spectra of light sources, is astronomy's fundamental interpretive tool.

A prism or grating spectrograph receives concentrated light from the telescope's objective on a narrow rectangular entrance slit. The light diverging past the slit enters a collimator, whose purpose is to deliver a beam of parallel rays to the dispersing device. After these rays of composite light either pass through a prism or reflect off a grating dispersing device, they will have been separated into their constituent wavelengths. The dispersed light is focused finally by a camera system on a radiation detector (a photographic plate or a photoelectric device) as individual color images of the entrance slit. Each wavelength forms a distinct image of the slit. The different color images of the slit are arrayed in an orderly progression of colors from red to violet to create a spectrum of the composite light.

Whereas the spectrograph is used to examine the spectral composition of radiation, the photometer is an analyzing instrument used to measure the amount of radiation coming from an astronomical object. It measures the amount of radiant energy on either a relative or an absolute scale at one wavelength or in a band of wavelengths. The radiation detector is generally today a photoelectric device, and thus the photoelectric photometer is much like an exposure meter on a camera: Incident light is converted proportionally into an electric current. One can use a variety of techniques to define the wavelength region for the photometer, such as color filters. Another way is to use the photometer in conjunction with a spectrograph, where the photometer is made to scan the spectrum formed by the spectrograph.

The photometer is usually limited to measuring only one light source, such as a star, at a time. But the limitation is compensated for by the photometer's very great accuracy. Because of its quick response to changes in amounts of light, a photoelectric photometer is particularly useful in continually monitoring the change in brightness of an object whose emission of radiant energy varies over time. For example, a number of stars are known to be variable light sources.

The Earth receives electromagnetic radiation of all wavelengths from various directions in space, but most of the electromagnetic spectrum is screened out by the atmosphere well above the Earth's surface. Wavelengths from only two regions of the electromagnetic spectrum, however, are able to penetrate the atmosphere to any extent. These two spectral windows in the atmosphere through which astronomers observe the Universe are called the optical window. This includes wavelengths from about 3000 to 10,000 A, or roughly the visible-wavelength region, and the radio window, which includes the wavelength region from about 1 mm to 30 m. The telescopes astronomers build to take advantage of these two atmospheric windows are thus logically called optical telescopes and radio telescopes. We discussed optical telescopes in the previous chapter, and we will consider radio telescopes in the next section of this chapter.

With the advent of the space age, astronomers have been able to use aircraft, balloons, rockets, and now primarily satellites to extend their vision of the Universe by going above part or all of the Earth's veiling atmosphere. Astronomers have been aghast at what the space-based telescopes have revealed through radiation in the ultraviolet, X-ray, gamma-ray, and infrared regions of the electromagnetic spectrum.

The theoretical resolving power of any optical telescope is never fully realized because the lower layers of our atmosphere are unsteady or turbulent, and turbulence blurs and distorts a star's image and causes it to twinkle, or scintillate. The rapid scintillations break the starlight into many dancing specks of light, which in long exposures merge to form the fuzzy stellar images we see in photographs. Even under the best conditions, optical images are no sharper than about 1 second of arc, that is, the angle subtended by a dime at a distance of 1 mile, and are typically several seconds across. The less the atmospheric turbulence, the less the stars twinkle, and the better is the seeing, as astronomers refer to it. A planet, however, shines with a steady light because each point on the tiny disk twinkles out of step with neighboring points; thus we see an average of all the twinkling points.

A technique called speckle photography can be used with large telescopes to get around the image smearing that comes from atmospheric turbulence. Several extremely short time exposures (less than 0.01 s) can be made, and in each the star image on the photographic plate appears as a cluster of sharp specks of different brightness. The information from each photograph can then be fed into a computer that reassembles the several photographic images into a single non-smeared image of the star.

Other nuisances hamper our observation of the heavens. The night sky's transparency varies as smog, dust, and atmospheric haze cloud it. The upper atmosphere is also suffused with a faint light called airglow. It arises when atmospheric molecules absorb ultraviolet photons from sunlight and then reradiate the energy in a few wavelengths of the green, red, and infrared spectral regions. On long exposures, airglow fogs a photograph and reduces the contrast between the faintest images and the sky background.

Another problem is that starlight penetrating the atmosphere is bent increasingly toward the vertical so that a star appears to be slightly closer to the zenith, that point directly above the observer, than it really is. This atmospheric refraction effect is greatest near the horizon, about $0.5°$, for there the light's path through the air is the longest. The consequences of this are that when we observe the rising or setting Sun, it is really below our horizon, since refraction has raised the Sun's image above the horizon by 0.5º, which is the Sun's own angular diameter.

5.4. RADIO TELESCOPES

In 1800, William Herschel detected the infrared component of solar radiation by positioning thermometers beyond the red end of the Sun's visible spectrum. His discovery foreshadowed the astronomy of those portions of the electromagnetic spectrum to which the human eye is not sensitive. What we have seen over the last 15 years in those spectral regions whose wavelengths are longer than the visible spectrum, namely infrared, microwave, and radio, has revolutionized our concept of the Universe. In this section, we discuss the detection of radiation in those long-wavelength regions.

A large portion of the infrared spectrum does not reach ground level because of absorption by water vapor, carbon dioxide, and molecular oxygen that lie between the ground and about 15 km of altitude. Consequently, airplanes, balloons, rockets, and satellites have been used to lift infrared telescopes above the veiling atmosphere. However, astronomers can also locate infrared observing facilities on mountaintops, such as the one in the Hawaiian Islands.

The modern infrared telescope is designed and functions much like a reflecting telescope used for visual observations with one major difference. Because bodies that are warmer than their surrounding, that is, contain more thermal energy than the surroundings, will attempt to equalize the energy content with those surroundings, they will radiate electromagnetic energy. If these bodies are only slightly warmer than their surroundings, say a few tens or hundreds of degrees Kelvin, they will radiate in the infrared portion of the spectrum and not the visible. Thus the major difference between a visual and an infrared reflector is that care must be given to the design so that the infrared reflector does not collect infrared radiation from warm objects nearby. Thus infrared reflectors have no tube around the telescope, have smaller secondary mirror supports, have been shield from hot electronic equipment, etc. The infrared detector, which generates an electrical signal when heated by infrared photons, works best when cooled to temperatures near absolute zero, such as with liquid helium. With a liquid-helium-cooled infrared detector on the appropriate analyzing instruments, the modern infrared telescope is a powerful research tool.

A number of modern telescopes have been designed and built for infrared astronomy only. A national observatory for infrared astronomy is located high on the 4200-m inactive Hawaiian volcano Mauna Kea. A 3.0-m infrared telescope constructed by NASA and the University of Hawaii is in operation there along with a 3.8-m infrared telescope belonging to the United Kingdom. Other major infrared telescope facilities are the Multiple-Mirror Telescope in Arizona, the University of Wyoming facility, and Mexico's 2.1-m reflector.

In 1931, a Bell Telephone engineer,

Karl Jansky (1905-1950), was trying to find where the interference disrupting transatlantic radiophone circuits came from. He discovered that some of the radio noise was not from the Earth, it was extraterrestrial. The primary source was the center of the Milky Way, in the constellation of Sagittarius. In 1936, an Illinois radio engineer, Grote Reber (1911-2002), pursued the phenomenon farther. He built the first parabolic radio telescope, 9.5 meters in diameter, and made the first radio map of the sky. The strongest signals he found came from the star clouds in Sagittarius and from several discrete sources toward the center of our Galaxy. The next major discovery was in 1942, by British radar operators and scientists tracking down suspected radar jamming during World War II; they discovered that the interference was radio emission from the Sun.

At first astronomers did not grasp just how significant Jansky's work was; they were preoccupied with their observations of the Universe through the optical window of the Earth's atmosphere. After World War II, however, radio astronomy came into its own when physicists, radio engineers, and astronomers joined forces to build larger and more efficient radio telescopes. Radio astronomy since then has led to startling discoveries, such as interstellar molecules, pulsars, and quasars. Today our concept of any celestial body is based upon its appearance all across the electromagnetic spectrum, and the radio region is an extremely important component.

Because the physical nature of a radio wave is exactly the same as that of a light wave, except for the longer wavelength, the problem of designing a radio telescope is similar in theory to that of designing an optical telescope. In practice, however, there are some differences. Radio waves pass through most materials without any interaction; thus it is not possible to design a "lens" for radio waves that will focus them in a refracting telescope. But any metal will reflect radio waves, so a dish-shaped metal mirror will focus radio waves, just as a glass mirror focuses light waves. The reflecting surface of the dish can be an open, fine-wire mesh or a solid metal with a parabolic shape. Radio waves are reflected from the surface and converge toward a focal point, where a small collector aerial absorbs the concentrated energy, turning it into an electric current. From there the current or signal is carried by an electrical cable to the receiving equipment, which processes the signal just as in your home radio receiver.

After amplification, the signal variations are recorded in one of several ways, with the final step being to feed the recorded signal into a computer for analysis. When the computer has done its job, a formerly invisible part of the Universe is revealed as a mapping of the intensity of the radio signal.

The sensitivity of radio telescopes can be increased, resulting in a greater pointing accuracy and higher resolving power, by enlarging the collecting area of the dish or by improving the capabilities of the receiver. With the largest radio telescopes we can obtain a resolution approaching a few seconds of arc, comparable with that of large optical telescopes. The most powerful radio telescopes can detect energy from sources whose power is comparable with that of a terrestrial FM broadcast station ten thousand billion km (10^{18} cm) away.

The radio telescope is remotely controlled by the astronomer from an electronic console, just as is the large optical telescope. Moderate-sized radio dishes, up to about 100 m or so in diameter, are steerable and have equatorial mountings that follow the rotating celestial sphere just as optical telescopes do. Larger and consequently heavier dishes, however, use an alt-azimuth mounting. This minimizes the distortion in the shape of the dish due to changing the orientation of the dish in the Earth's gravitational field.

Even larger and more unwieldy antennas are fixed, pointing upward always, while the rotating Earth sweeps much of the sky by the antenna's field of view. Arecibo, Puerto Rico has the biggest fixed antenna, a metal dish 305 m across contoured out of a natural bowl in the ground. It can survey the sky to within 20° of the zenith, allowing coverage of about 40 percent of the entire sky.

Astronomers have searched endlessly for better resolving power. It can be achieved by building bigger telescopes or by observing at shorter wavelengths for a chosen aperture or by using the phenomenon of interference discussed previously. Radio interferometry is a technique involving two or more radio telescopes. Optical telescopes have also been linked in a similar manner and such an arrangement is called optical interferometry. Radiation from an astronomical source received at the individual telescopes is combined to obtain data that have a spatial resolving power equal to that of a single telescope as large as the distance between the individual receivers. With radio interferometry an astronomer can obtain details about the spatial structure of a given celestial object that a single radio telescope could never reveal.

The separation between the individual telescopes of a radio interferometer is limited only by the ability to correlate the results from each radio telescope, because the technique depends on combining, at the same instant, the signals received by the separate telescopes. With the advent of the atomic clock, that is, a clock governed by the vibrations of certain atoms, it became possible to record the signals received by the different telescopes, along with the precise time, and to compare them later. This allowed the individual telescopes to be greatly separated, even on opposite sides of the Earth. The technique is called very long baseline interferometry (VLBI).

In the Very Large Array (VLA) radio interferometer recently put into operation in New Mexico signals from each of 27 individual radio telescopes are combined by a computer. Each dish is 25 m in diameter, and the 27 individual telescopes are moved along railroad tracks arranged in the shape of an enormous 21-km Y. Nine dishes can be located on each branch of the Y, and the system can provide a total of 351 interferometer pairs of antennas. The energy-collecting power of the VLA is roughly equivalent to a single 122-m telescope, and it has a spatial resolution smaller than 1 second of arc or better than that of the 5-m Hale optical telescope.

A proposal has been made that the VLA become the heart of an array of ten 25-m radio telescopes widely spread from Hawaii across the United States to Puerto Rico. Known as the Very Long Baseline Array

(VLBA), it would operate as a single instrument, with each antenna directly controlled from the main operation center in Socorro, New Mexico. This would give the VLBA the ability to ferret out the structure of radio sources as small as 0.0003 second of arc in diameter. Although it would operate as primarily a ground-based facility, a proposal has also been made to complement VLBA by putting a single large antenna in Earth orbit to work with it, making a truly immense radio interferometer.

Chapter 6
The Earth and Moon

The first five chapters laid a foundation for the exploration of the Solar System. One important aspect of that foundation was the consideration of the structure or processes of science. Our discussion focused on first, the terms in which a scientific argument may be cast, second, the underlying preconceptions or themes in science, and third, what constitutes a scientific explanation of our experiences with such natural phenomena as motion and light. Scientific explanation is expressed in Newton's theory of motion and in Maxwell's, Planck's, Einstein's, and Bohr's theory of a dual nature for light. Although Newton's theory of motion is not the final word on the subject of motion, for we have yet to consider Einstein's relativity concept, Newtonian theory has been abundantly successful in predicting what has been found regarding motion in the Solar System all the way from the motion of planets to the launching of spacecraft to survey distant parts of the system. Likewise, the wave and photon concepts of light provide an abundant basis for interpreting light coming either directly to the Earth or first to a wandering spacecraft which relays its finding on to the Earth. From the study of light we have gained an understanding of the physical nature of self-luminous sources like the Sun and reflecting bodies such as planets.

The Solar System is an awkward mixture of bodies. The Sun is a star, a very different kind of body from planets, or satellites, or asteroids, or comets. Even the planets are not simply larger or smaller versions of the same type of body. Consequently, since the Sun is the only star in the Solar System, we shall delay its consideration until later. In the following chapters, we will touch upon the other occupants of the Solar System, asking what they are like, how they do or do not resemble each other, reasons for their being as they are, and what reasonable scenarios exist to account for their evolution from primal matter to the bodies we recognize today.

As long as humanity has been aware of the sky, the five brightest objects after the

Sun and Moon, were known to be different from the remainder of the stars. These are now named Mercury, Venus, Mars, Jupiter, and Saturn. Galileo realized that the Earth was but another of the Sun's planets. The number of known planets did not change until the use of the telescope, after which Uranus was discovered in 1781, Neptune in 1846, and Pluto in 1930. The discovery of the Solar System's minor bodies, also primarily after the invention of the telescope, broadened our conceptual understanding of the Solar System from the Greek's concept of seven wandering bodies to a far more complex set of objects with relationships existing among them that had ever before been envision. By the turn of the 20th century, the full extent of the Solar System was beginning to be revealed.

Astronomers now use aircraft, balloons, rockets, and now primarily satellites and orbiting observatories to extend their understanding of the Solar System by going above part or all of the Earth's veiling atmosphere. Amazed at what the space-based telescopes have revealed through radiation in the ultraviolet, X-ray, gamma-ray, and infrared regions, astronomers now vigorously pursue knowledge with more advanced spacecraft.

The dominant body in the Solar System is the Sun. It is the source of the gravity that bonds the Solar System together. The Sun is also the source of radiant energy that powers so many physical processes going on throughout the Solar System. However, the Sun is a star and is therefore quite unlike any other Solar System body. Like other stars, the Sun is a gas from center to surface, possessing a radius over 100 times greater and a mass over 300,000 times greater than that of Earth. The Sun generates deep within its hot interior the radiant energy that it radiates from its surface. The Sun's family of planets intercepts only a minute fraction of this radiation flooding the Solar System.

In addition to the steady emission of radiant energy, there are numerous transient phenomena occurring in the Sun's atmosphere, such as sunspots, flares, and prominences. Associated with these is a flow of subatomic particles and magnetic fields out through the orbital planes of the planets. This solar wind of particles impinges on the planets and their magnetic fields, producing a variety of phenomena, such as Earth's aurora (northern and southern lights). Thus the range of interaction between the Sun and its planets is far more complex than just gravitational.

After the Sun, the eight major planets as a group contain the next largest fraction of the Solar System's mass. The variation in mass among the nine planets, however, is immense with Jupiter being over 300 times the mass of our Earth.

Although we believe the planets had a common origin in time, some 4.6 billion years ago, they currently display significant chemical, physical, and geologic differences. Such diversity stems primarily from their different masses and distances from the Sun at the time of their formation. During that formative period, these factors determined the ability of the fledgling planet to retain matter and further they defined the chemical composition of that matter.

In spite of their differences, we know of sufficient chemical and physical similarities among planets so that we can divide them into two categories, the Terrestrial planets and the Jovian planets. The Terrestrial planets, consisting of Mercury, Venus, Earth, and Mars, occupy the inner Solar System and are composed mostly of iron and its oxides and silicates. The Jovian planets consisting of Jupiter, Saturn, Uranus, and Neptune, define the outer Solar System with all being farther from the Sun than the Terrestrial planets. They are also larger than the Terrestrials and are composed primarily of the light elements hydrogen and helium; the most abundant elements in the Universe. Jupiter and Saturn apparently have a chemical composition somewhat like that of the Sun, while Uranus and Neptune seem to have relatively more carbon, nitrogen, and oxygen than hydrogen and helium. The important point is not to mentally visualize the planets as being either the same size or type of bodies for they are very far from being a category of uniform bodies.

Satellites or moons of the two planetary groups are as distinctive in their physical structure as are their parent groups. There is also variation in the sizes of the objects that we label as moons. For example two of Jupiter's moons, Ganymede and Callisto, and one of Saturn's satellites, Titan, are as large as, or larger, than Mercury. Our Moon is not all that much smaller than Mercury either. Thus it is not size or physical similarities that define moons as a category, but it is their relationship to bodies we identify as planets, along with the historical tradition of having always done so.

One of the most exciting developments in planetary research has been the discovery of ring systems for all the Jovian planets. Rings are actually individual, small solid bodies in orbit about a planet usually in its equatorial plane. They are thus very small satellites of the planet.

Unlike planets and their satellites, comets are icy fragments impregnated with rocky matter coming from deep space well beyond the orbits of the planets. Nevertheless, they are members of the Sun's family of Solar System bodies. Their icy composition is apparently characteristic of many bodies in the outer Solar System, including most of the Jovian satellites. As comets come into the inner Solar System drawn by the Sun, they evaporate in the intense solar radiation strewing rocky bits and pieces along their orbital path. This material forms much of the meteoric material.

Asteroids are small, rocky bodies which display as much diversity in size as do planets and satellites. And those asteroids in the inner Solar System near the orbit of Mars which have been study for chemical composition, suggest again a variation such that there is no close uniformity in structure among them either. In recent years, the term asteroid has been expanded to include small objects, presumably not comets, located in the outer Solar System. It is unlikely, however, that their physical makeup is like that of asteroids in the inner Solar System.

Meteoroids range in size from irregular solid bodies, called meteorites when they strike the Earth's surface, to tiny particles, called meteors if they merely flash through our atmosphere. As we go down the scale in size, the number of meteoroids increases rapidly. All the meteoroids are satellites of the Sun and are moving in orbits that vary widely in their characteristics. They are composed of rocky material and are apparently derived primarily from asteroids and comets.

The interplanetary medium is composed primarily of gas particles, mostly protons and electrons that are ejected from the Sun's atmosphere. These subatomic particles form the solar wind mentioned above. Some dust is there too, most being cometary debris. Despite huge numbers of gas and dust particles, interplanetary space has fewer bits of matter and is a better vacuum than can be made in a terrestrial laboratory.

Professional astronomers do very little naked-eye study with telescopes of planets or the minor Solar System bodies. The primary use of a telescope, regardless of whether it is here on Earth or located in a spacecraft, is as a camera for taking direct photographs, or for recording spectra, or for photometric measurements. This is because these techniques produce permanent records that can be studied as needed. In addition, using radiation detectors such as photographic emulsions or photoelectric devices makes quantitative studies possible, something generally not possible with naked-eye observations. As a reminder, photometry means making measurements of the amount of radiant energy a body gives off, while spectroscopy is the analysis of light by separating composite light into its component colors.

Photometric measurements provide information about the nature of reflecting materials, such as clouds in a planet's atmosphere or surface features, spectroscopy provide clues to the chemical composition of a planet's surface, clouds, and atmosphere. Using the spectrum of sunlight reflected from the planet by either its atmosphere or surface, may reveal a planet's atmospheric constituents by the absorption lines or bands that are superimposed on the solar spectrum.

In studying the radiation from planets, or for that matter and astronomical body, we divide the radiation into two categories, thermal and non-thermal radiation. Thermal radiation, which is due to the fact that a body is hot, can be studied from the ultraviolet to the far infrared region with today's modern radiation detectors. Thermal radiation is blackbody radiation which possesses a continuous spectrum and is the product of the random thermal motion of particles that compose the outer portions of planets, satellites, asteroids, or any other astronomical body. Because of the very low temperature of planets compared to stars, thermal radiation coming from a planet is situated primarily in the infrared portion of the electromagnetic spectrum. Such infrared data provides important information on surface and atmospheric temperatures and, indirectly, chemical composition.

Non-thermal radiation is radiation due to physical processes other than that involved in producing thermal radiation. That is, it owes its explanation to some other fact than that the body is hot. For example, light produced in lightning is non-thermal radiation. Both thermal radiation and non-thermal radiation emitted by planets can in some cases be observed with radio telescopes.

Since we cannot directly determine the chemical composition of the deep interior of even the Earth, how is it possible to think that we know the composition of planets and other Solar System bodies? Above we mentioned ways of obtaining evidence as to the compositions of atmospheres and surface layers, but that is far from believing one knows the composition throughout the body. As you already know, the Earth's atmosphere is made up of nitrogen and oxygen, while the surface layers of the Earth are rich in silicon, oxygen, and aluminum. Thus is it possible that the main body of the Earth may have a composition that is different from either its atmosphere or surface? We believe that, yes, it is indeed different, and we have arrived at that conclusion not by means of experimentation, although some types of experimental results are important to the argument, but primarily by means of a theoretical argument. The argument proceeds along the following lines.

Whatever composition the planets had at their birth has clearly not changed over the span of their lives, since no significant influx of new material from outside has been added to the planets' masses and no known processes are at work to change their original composition into something else, as there are in the case of stars.

What then was the original composition of the planets, and is it likely that it is not the same for each planet? As the Solar

System formed, it is probable that the Sun was well along in the process and consequently reasonably hot by the time the planets began to form. Thus, the temperature of the matter from which the planets formed, was higher, closer to the Sun, probably 2000K or so, and declined rapidly outward to about 100 K. Since matter should be solid or solid-like to coalesce to form a planet, we can divide the chemical compounds most probably present at the time of the formation of the Solar System into three broad groups on the basis of the ease with which they vaporize.

The first group, called the gaseous materials, consists of those elements that are gases at temperatures above a few tens of Kelvins, such as hydrogen and helium. Next are the icy materials such as methane, ammonia, carbon dioxide, and water, containing such light elements as carbon, nitrogen, and oxygen besides hydrogen, which do not vaporize until the temperature is over a couple of hundred degrees. Finally, there are rocky materials, such as iron, magnesium, and their oxides, sulfides, and silicates, which remain solids until the temperature is several thousand degrees. Hence close to the Sun, where the Terrestrial planets formed, the rocky materials would have been the only ones not in a gaseous form. Iron and the elements near it in the periodic table should dominate the compositions of the Terrestrial planets, as they seem to do, while lighter elements, such as hydrogen, helium, carbon, nitrogen, and oxygen, should be the principal constituents of the Jovian planets, as they seem to be. This form of argument in science has proven to be most fruitful since it provides many opportunities to predict and then to verify or not those predictions through observations.

From space, the Earth's appearance to the eye is that of a pale blue dot, as highlighted by Carl Sagan (1934-1996). Swirling and moving through the atmosphere, just barely above the surface, are white fluffy clouds. From any distance, say the Moon, the Earth appears to the eye to be quite spherical with an absence of any surface relief. It is not until one approaches much closer to the Earth that any height variations become apparent.

Our nearest cosmic neighbor, the Moon, has a different appearance from space than that of Earth. There is no watery surface, no white fluffy clouds floating over the surface. Like Earth though, the roughness of the lunar surface is not apparent to the eye until one is actually quite close to it. From close to the lunar surface, it is obvious that the cratered surface bears little resemblance to the active and ever-changing surface of Earth.

One of the most significant concepts in our view of the physical world, although it began in ancient Greece, did not become a dominant consideration in scientific thought until the eighteenth century. Within a period of about 100 years it revolutionized scientific thinking. What was that concept? It is the concept of evolution. Although you may have encountered the concept in connection with biological evolution, it is a much broader one than just biology. Today our world view is one dominated by a knowledge that the Universe, including galaxies, stars, the Earth, and life on Earth's surface, is gradually, sometimes rapidly, evolving in directions shaped by natural processes governed by the laws of physics. Throughout this chapter and those which follow we shall try to sketch not only the results of evolution but also its processes.

Let us begin our study of the planets of the Solar System by considering the distinctive differences between the Earth and the Moon. First we need to describe what the two bodies are like and then consider in the next chapter what evolutionary process have occurred over their life times that has made these two bodies, that are so close to each other, so dramatically different.

6.1. OUR HOME, THE EARTH'S SURFACE

Although from space the Earth may appear to be spherical, in point of fact it is not. Measurements at different points over the surface reveal that the number of kilometers in 1° of latitude increases slightly from the equator toward the poles. This means that the Earth's shape is actually that of an oblate spheroid with the longer diameter in the equatorial plane and the shorter one in the polar direction.

Earth's rotation is primarily responsible for causing the departure from the true shape of a sphere. Rotation causes the body of the Earth to flow from high latitudes toward lower latitudes, forming an equatorial bulge. Because Earth is not a perfect sphere, its gravitational field is not the same in all directions and such variations affect the motions of artificial satellites. From these unanticipated changes in satellite motions, called orbital perturbations, we can work the problem backward to find the Earth's shape, which we now know is slightly pear-shaped. The stem portion at the North Pole is about 19 m farther from the center, and the bottom portion at the South Pole about 26 m closer to the center.

Several phenomena were used historically to demonstrate the rotation of the Earth besides the rising and setting of the Sun and stars. One of the most vivid was the pendulum experiment devised in 1851 by the French physicist Jean Foucault (1819-1868). He hung an iron ball on a long wire from the dome of the Pantheon in Paris. Underneath it was a large circular table with a ridge of sand along its edge. As the pendulum swung, a pin attached to the bottom of the ball made a mark in the sand.

After the pendulum had been set into motion, it was apparent from the marks in the sand that it was deviating slowly in a clockwise direction. From Newton's laws of motion, we know that once a plane of oscillation has been established for a pendulum, only an external force can change the plane's orientation. No external forces were acting on Foucault's pendulum; in reality it was the spectators and building that were turning underneath the plane of oscillation of the pendulum because of the Earth's rotation.

Consider an idealized Earth whose surface is entirely covered by water. Since the gravitational force exerted by the Moon decreases as one over distance squared, the Moon's attraction is greatest for the ocean closest to the Moon, and it decreases across the Earth, so that it is least on the ocean on the far side opposite the Moon. Relative to the Earth's center, this difference in force across the body of the Earth, called a tidal force, causes the ocean on the side nearest to the Moon to shift slightly closer to the Moon and the ocean on the opposite side to recede slightly. That is, the Moon's gravitational pull on the oceans produces two tidal bulges on opposite sides of the Earth in line with the Moon. Consequently, water piles up in the form of an ellipsoid whose long axis is directed toward the Moon. Midway between the high tides are the low tides.

The Earth's rotation underneath the tidal bulges results in alternating high and low tides in the oceans twice each day. Because there is a slight lag before the oceans fully adjust to the Moon's tidal force, the tidal bulges are dragged by the rotating Earth somewhat ahead of the line joining the centers of the Moon and the Earth.

The Sun also contributes to the tides, but only half as much as the Moon does because of its much greater distance, despite its larger mass. When the Sun and the Moon are roughly along a straight line, such as at new or full moon, their combined gravitational pull is greatest, producing the largest tides.

If the Sun and the Moon are pulling the Earth, why doesn't the land move too? It does, because the land is not absolutely rigid. The land has a greater internal strength than water, and therefore, land tides are much smaller than water tides. Approximately every 12 hours the ground on which you stand rises and falls a few centimeters at any given place. Tidal motions are also evident in the atmosphere, which is even less rigid than the oceans.

The constant friction generated by the lunisolar tides, mainly near the shores and in the shallow seas, has slowed the Earth's rotation. As a result, the day has lengthened over several billion years by an estimated several hours to the present 24 hours. The slowing down in the rotation is not uniform; a number of irregularities have been found. This conversion of Earth's rotational energy into heat by tidal friction will continue indefinitely.

The outer layers of the Earth consist of a crust and the uppermost part of what is called the mantle, and together they are known as the lithosphere. This is a fairly rigid zone of rocks that extends about 100 km below the surface in which the crust extends some 60 km or so beneath continents, but only about 10 km below the ocean floor. Rocks composing the continental crust have a lower density than those composing the oceanic crust, and they are primarily a light-colored granite which is rich in the silicates of aluminum, iron, and magnesium. In a simplified view, the continental crust possesses a layered structure: On the bottom is a layer of igneous rock, that is, molten rock that has hardened, such as granite, over which lies a thin layer of sedimentary rocks, those formed by sediment and fragments that water deposited, such as limestone and sandstone. Overlying the sedimentary rocks is a layer of soil that has been deposited during past ages in those parts of continents that have not recently experienced volcanic or mountain building activity.

Sandwiched between the lithosphere and the lower mantle is a 150-km thick layer of partially molten material. This layer, called the asthenosphere, is readily deformed and can be made to flow when pressure is exerted. Its chemical composition is primarily iron and magnesium silicates.

In efforts to date various regions of

continents, geochemists have shown with radioactive dating techniques that the oldest rock formations on continents have ages between 3.5 and 3.8 billion years. For North America, the oldest part is a crescent-shaped region bordering the west and south sides of Hudson's Bay. A younger crescent lying roughly to the west and south surrounds this oldest region, and the westernmost and southernmost parts of the continent are even younger. A somewhat similar pattern exists for other continents. The inference is that continents are not original with the Earth's formation 4.6 billion years ago but are a secondary aspect and they will continue to grow and evolve. We know that continental margins, particularly the western edge of North America, are new additions to continents. These coastal regions are growing due to the deposition of sediments washed down by rivers from the interior of the continent. In striking contrast, the oldest known parts of the oceanic crust are about 200 million years old or almost 20 times younger than the oldest parts of continents.

The concept that continents move or drift relative to each other was not one that was readily accepted when the German geologist Alfred Wegener (1880-1930) proposed it in 1912. Yet recent research has revealed a variety of evidence showing that the lithosphere is indeed segmented into about a dozen or so major plates of different sizes. Floating on the mantle, these plates move slowly, carrying the continents with them at a typical rate of several centimeters each year. This motion is known as plate tectonics or, more popularly, as continental drift.

One type of evidence for lithospheric plates comes from extensive exploration of the ocean floor which has revealed the existence of a number of mid-ocean ridges rising several kilometers above the ocean floor and extending thousands of kilometers in length. We now know that these ridges mark one type of plate boundary. Lithospheric plates are internally quite rigid so that their principal interactions with each other are on their boundaries. As a consequence, the boundaries are the locations of large-scale geological activity. An example of a mid-ocean ridge is the Mid-Atlantic Ridge that separates the North and South American plates from the Eurasian and African plates. For the mid-ocean ridge-type of boundary, it appears that lava is first being forced upward from the asthenosphere into the ridges from which the lava pushes out laterally from the ridge. This new material gradually cools, thickens, and solidifies at the trailing edge of the plate. Rock samples from as far down as 8 km below sea level verify that the Earth's youngest volcanic rocks are those found near these mid-ocean ridges.

We have further confirmation that the plates move from the shape, geologic structure, and fossil record of continents. Evidence also comes from igneous rocks with similar magnetic fields that were frozen into the rocks at the time they solidified. Such rocks have been found at continental margins that are now widely separated from each other.

Another line of evidence comes from heat flow out of the interior. Compared to the energy falling on the Earth from the Sun, the interior flow is scarcely a trickle: The heat conducted through an area the size of a football field is roughly equivalent to the energy given off by three 100-watt light bulbs. Yet over Earth's 4.6-billion year history this trickle of energy has contributed to the work of making continents drift, opening and closing ocean basins, building mountains, and causing volcanoes and earthquakes. The geographic variation in the heat flow from the interior is not great, but the global variation shows that the major oceanic ridges are high-heat-flow zones, while the older continental shields and sedimentary regions are low-heat-flow zones.

How are lithospheric plates transported across the top of the mantle? It appears that they are driven by a horizontal flow of convective currents in the upper, softer portion of the mantle. Convection is a process that transports energy from one place to another. The closest example to our everyday experience is the rising of heated air in a room followed by cooler air sinking to the floor to be reheated and cycle upward again. If new plate material is being added at a plate's trailing edge near a mid-ocean ridge, then other material must be taken off the plate at some other location. This process is occurring at the leading edge of one plate that is being pushed downward underneath an overriding plate to create a deep ocean trench where the two plates meet. Such a process forces old plate material into the mantle to melt and be recycled over hundreds of millions of years as part of the convective currents that drive the plate motion. This process can form a coastal mountain belt, like the Andes, on the overriding plate. Over millions of years as the other plate descends, it heats up and becomes part of the general circulation in the asthenosphere. Plates separate along mid-ocean ridges. Most of the great geologic processes, such as volcanic activity, mountain building, formation of ocean trenches, earthquakes, are concentrated near plate boundaries.

About 200 million years ago, the last mass movement of continents began. There was at that time but one single land mass, today called Pangaea. This supercontinent probably accumulated from migrations produced by previous drifting. Some 20 million years later, sea-floor spreading had separated the supercontinent into two segments. About 45 million years later, the North Atlantic and Indian Oceans widened and South America began to separate from Africa. During the next 70 million years, the South Atlantic Ocean widened into a major ocean, the Mediterranean Sea began to open up, and North America just began to separate from Eurasia. A computer-generated projection for the next 50 million years suggests that the Atlantic and Indian Oceans will enlarge and the Pacific will contract. Africa's northward movement will eventually doom the Mediterranean.

Although typical plate motions are a few centimeters per year, this reshaping of the Earth's face is actually quite dramatic when one considers the age of the Earth. It is estimated that in about 2 billion years the gradual cooling of Earth from heat loss will mean that the asthenosphere will flow less readily and the plate-motion phase of the Earth's evolution will probably come to an end. Thus the Earth will enter a new phase, in which plate motions of the lithosphere are not responsible for most of the large-scale terrain features. Large mountain ranges, like the Himalayas, will no longer be uplifted, and they will erode away over millions of years.

As far as surface geography is concerned, there appear to have been two major terrain-shaping mechanisms at work on the Earth and, for that matter, presumably on the Moon, Mars, Venus, and Mercury, the other Terrestrial planets. These are impact cratering, whose most intense period of bombardment some 3 billion years ago is now long past, and thermal-tectonic activity due to an outflow of thermal energy from the deep interior. The thermal-tectonic mechanism, that is, the plate motions and deformations of the crust with accompanying volcanic and earthquake activity, aided by erosion by wind, water, and life processes, is now the dominant one, but only on Earth. It has all but erased the results of that long-past impact cratering phase in Earth's history, except that remnants of the last of that phase can still be seen in nearly a hundred ancient impact structures, some of which are as large as the largest visible ones on the Moon.

It is estimated that on Earth thermal-tectonic activity is responsible for better than 90 percent of the present terrain, with not more than 10 percent of the cratered terrain remaining. Present evidence suggests that surface evolution on the other terrestrial planets, as revealed by various space missions, has not been so heavily dominated by thermal-tectonic activity as that for Earth.

A revolution is in the making in the study of the geography of the surface. Satellites now orbit the Earth photographing the surface in narrow-wavelength regions that are in either the visible or the near infrared. Coverage of each photograph is about 30,000 km^2 with resolutions varying from about 100 m down to 10 m. These satellite images coupled with the power of the modern supercomputer to analyze data from them promises to enhance geographical and ecological studies beyond the boldest expectations of just ten years ago. Earth surveillance will dramatically change agriculture, prospecting for natural resources, fishing, economic development, commodities trading, and a host of areas concerned with human activities. In the next century little will happen on the surface, such as a developing crop failure in Africa, or an oil pipeline break in the Middle East, or the movement of cargo vessels at sea, about which the information will not be almost immediately available worldwide. It is even hard to visualize all of the uses to which this constant surveillance data will be put.

6.2. EARTH'S INTERIOR

Although the surface is far from perfectly understood, its study at least holds out the hope that we may go where necessary to either directly observe or perform some type of experiments that will improve our understanding. The interior of the Earth, however, is a far different situation in that at no time will we venture to the center to observe its physical processes. The internal structure of the Earth has and will continue to be a problem that we must attack from afar. Knowing what happens on the outer boundary of the region we wish to investigate, such as measuring the heat outflow from the interior, is a start. But what we really need is a probe that reaches into the deepest parts of the body of the Earth to reveal its secrets. And conveniently enough, Earth itself provides us with that probe.

From what we know about the rotation of the Earth, it is apparent that the Earth is not absolutely rigid. The body of the Earth will deform when subjected to various forces, such as those occurring in the propagation of waves. In such waves, known as seismic waves, geophysicists have a natural probe of the planet's deep interior. Seismic waves are generated by earthquakes and spread out in all directions from the site of the quake. There are also artificial means of producing seismic waves, such as explosives. From the manner in which these waves propagate throughout the body of the Earth, i.e., their periods and amplitudes of vibration, and their arrival time at various observing stations, we can deduce much about Earth's internal structure.

There are two kinds of seismic waves, pressure (P) and shear (S) waves, that propagate inside the Earth. The speed with which these waves travel through the Earth, between 5 and 15 km/s, depends on the material's density, compressibility, and rigidity. The particles of the Earth that transmit the P waves vibrate back and forth in the direction in which the wave propagates, similar to the way sound waves are propagated through air. The S waves cause the particles that transport the disturbance to vibrate perpendicular to the direction of the waves' propagation, as waves on a string do. S waves move at about half the speed of the P waves. Also unlike P waves, S waves cannot propagate through liquids, which damp their vibrations. As the P and S waves move downward through the Earth, their speed increases with the increasing density of the material they are traversing. Since seismic waves are refracted or reflected on reaching a boundary between two distinctly different layers, a picture of the Earth's interior can be produce by tracking their path through the body of the Earth. Such pictures show that the Earth possesses a layered structure like that of an onion.

All models of the Earth's internal structure are derived from seismic studies. At the center is a hot, high-density inner core, presumably solid and composed of iron and nickel primarily. We infer from the inability of S waves to penetrate the core to any extent that surrounding the inner core is a liquid outer core. This outer shell of molten material is also composed of iron and nickel, primarily with whatever lighter materials are present, being able to rise to its top.

Surrounding the core is an envelope known as the mantle. The upper portion of the mantle is mostly solid rock composed of olivine, an iron-magnesium silicate, while the lower portion is chiefly iron and magnesium oxides. Overlying the mantle is a thin crust of metal silicates and oxides, such as basalt, which is largely oxygen, silicon, aluminum, magnesium, and iron, and granite, which is oxygen, silicon, aluminum, sodium, and potassium. Obviously, since we live on top of the crust, we have the most immediate knowledge about its structure and composition. The degree of certainty of our knowledge of the interior diminishes deeper down toward the Earth's center.

Earth's internal structure gives all of the indications that it is a product of a completely molten body having undergone chemical differentiation, which is the process of heavier materials settling under gravitational attraction, while lighter materials migrate outward. This molten period when chemical differentiation took place seems to have been shortly after the Earth's formation.

6.3. EARTH'S ATMOSPHERE

If the Earth had no atmosphere, life would not and could not exist here. The insulating blanket of air surrounding us maintains a temperature range favorable for life in part because the incoming solar radiant energy, which is primarily in the visible wavelengths, is trapped by atmospheric carbon dioxide and water vapor molecules. What they do is to absorb energy coming from the surface of the Earth and reradiate much of it back to the surface. This process is known as the greenhouse gas effect. Without its atmosphere the Earth's average temperature would be about 20 to 30 degrees lower than its present value of 15° C (288 K). Since water freezes at this temperature, it could neither have facilitated the development nor could it maintain life as it does in its liquid form. Worldwide circulation of the atmosphere also transports thermal energy and helps to moderate extremes in temperature that would otherwise exist.

Moreover, even the upper atmosphere is important for our survival. It protects us from harmful ultraviolet and X-ray radiation from the Sun, vaporizes meteoroids entering the atmosphere, and absorbs most of the incoming highly energetic subatomic particles called cosmic rays. Finally, the atmosphere creates the soft blue appearance of the sky. This is because atmospheric gases scatter photons in the blue region of incoming sunlight much more efficiently than they do photons of longer wavelengths. This is why the rising or setting Sun is redder than when highest in the sky. There being more atmosphere along the line of sight toward the horizon than along the line of sight towards the zenith, more blue photons have been deviated in directions away from our line of sight toward the horizon than toward the zenith.

In all, the Earth's atmosphere plays a

very vital role beyond the obvious one of providing the oxygen we breathe. One of humanity's most important challenges is to understand and to preserve the atmosphere, for our continued existence depends on that knowledge.

The mass of the atmosphere is about one-millionth the total mass of the Earth. It has several layers each with distinctive thermal, physical, chemical, and electrical properties. Approximately half the atmosphere is contained in the first 5.6 km, and 99 percent of it lies below 35 km.

Our weather takes place in the bottom layer, called the troposphere. At an altitude of 11 km, the temperature drops to -55° C. Above this region lies a 40 km thick layer, the stratosphere, where the temperature slowly rises, reaching a maximum of about 0° C at 50 km. Somewhat below this altitude an absorbing layer of ozone screens out most of the incoming ultraviolet radiation. Within the next layer up, the mesosphere, the temperature rapidly drops to a minimum of -85° C at its upper limit, which is 90 km.

Above the mesosphere is the thermosphere. Here the still more dangerous

X-rays and gamma rays are effectively filtered out by molecular oxygen and nitrogen and by their dissociated atoms at even higher altitudes. The temperature climbs steadily throughout the thermosphere and into the exosphere, the atmospheric fringe several hundred kilometers above sea level.

Within the Earth's atmosphere are layers in which the concentration of free electrons is above the average atmospheric value. These layers constitute the ionosphere. The electrons are due to the ionization of atmospheric molecules and atoms by solar ultraviolet and X-ray photons. Radio waves of certain wavelengths, for example, the AM band of conventional broadcasting, transmitted by ground stations are reflected between the ionosphere and the Earth's surface. This makes possible long-distance communication between stations that are not along a direct line of sight because of the Earth's curvature. Radio wavelengths greater than about 10 m are turned back by the ionospheric layers, whereas shorter wavelengths pass through the ionosphere into space with little or no bending.

Up to about 90 km, gravitational settling causes no significant separation of atmospheric gases by atomic weight. No separation occurs because the atomic and molecular constituents are mixed by air currents and random thermal motion. The chemical composition of the atmosphere therefore remains nearly uniform, with 77 percent nitrogen, 21 percent oxygen, nearly 1 percent argon, 0.03 percent carbon dioxide, and almost 1 percent water vapor (which varies up to several percent in the troposphere). The atmosphere has minute traces of other gases, including neon, krypton, xenon, methane, ammonia, nitrous oxide, carbon monoxide, and ozone.

Above about 90 km or so, the constituents are not well mixed; the heavier molecules and atoms settle toward the bottom, the lighter ones diffusing to the top. At extreme heights a rarefied layer of helium extends from about 600 to 1000 km, and this is topped by a very tenuous hydrogen layer that merges into interplanetary space.

The chemical composition of the atmosphere is not static. The present composition results from a balance between those processes which introduce a particular molecule into the atmosphere and those which remove it. Probably the most significant example is that of oxygen, since it is essential for our existence. Atmospheric oxygen is almost entirely produced in photosynthesis, primarily by green plants in shallow seas and to a lesser extent by plant life on land. A little oxygen comes from the direct dissociation of atmospheric water molecules by ultraviolet photons from the Sun. Chemically, oxygen is quite an active molecule, combining readily with a number of different atoms, such as in the formation of oxides in rocks removing oxygen from the atmosphere. Breathing by animal life also depletes atmospheric oxygen. If the supply of oxygen were shut off, it would take only a few tens of thousands of years to remove the major portion of oxygen now existing in the atmosphere.

The abundance of the other molecules in the atmosphere is also controlled by various "production and destruction" processes. And as in the evolution of the Earth's surface, the atmosphere has also changed over time.

Clearly, if atmospheric oxygen is due to the existence of life, then oxygen would not have been present prior to the emergence of life. The origin of the primitive Earth's atmosphere is probably the result of outgassing by volcanoes and the escape of gases from the crust. The gaseous emission from present-day volcanoes includes water vapor, carbon dioxide, nitrogen, inert gases, and small amounts of methane, ammonia, and sulfur compounds. It is probable that on the very young and lifeless Earth, with no significant amounts of liquid water, the dominant atmospheric constituent was carbon dioxide in a very dense atmosphere. This estimate is based on the fact that a large amount of carbon dioxide is trapped in carbonate rocks on the Earth's surface. Carbon dioxide is the main component of the atmospheres of Venus and Mars, with the Venusian atmosphere being some hundred times denser than ours. About 2 billion years ago the transition began to an oxygen-nitrogen atmosphere. The amount of oxygen grew from a trace to the present 21 percent as a result of the development of photosynthesis by green plants, while carbon dioxide diminished.

6.4. EARTH'S MAGNETOSPHERE

That Earth possesses a magnetic field is not a new discovery for the compass, which is responding to the magnetic field, has been in use for some time. When it became known that the Earth's interior is hot, however, it was obvious that the Earth's magnetic field could not be like a permanent magnet. This is because heating disorients various parts of a magnet destroying its ability to produce a coherent magnetic field. Thus arose a puzzle as to where the Earth's magnetic field comes from. Scientists now know it to be caused by the complex circulation of liquid metal in the outer core: If friction can ionize metal atoms, then the flow of ionized material becomes an electric current, which produces the magnetic field. Such a mechanism is known as a dynamo, a device that converts mechanical energy of motion into electrical energy. Thus the Earth is more of an electromagnet than a permanent magnet.

In appearance the Earth's magnetic field

resembles that of a bar magnet inclined slightly to the Earth's axis of rotation. The magnetic field lines run between northern and southern polar regions, much as the pattern formed by iron filings sprinkled around a bar magnet. The intensity of the magnetic field decreases away from the Earth's surface, but the magnetic field can still be measured many tens of thousands of kilometers out in space.

However it began, Earth's magnetic field has changed polarity, that is, the north magnetic pole becomes the south magnetic pole and vice versa, many times over geologic history. Scientists trace the record of these changes through the magnetism frozen into rocks of different ages: Iron particles in molten lava beds align themselves along the lines of the existing magnetic field lines, and after the rocks solidify, they retain that orientation indefinitely. Such rocks show that magnetic reversals have come at intervals as short as 35,000 years. Why the reversals? One suggestion is that they are related to changes in the Earth's rotation or in the fluid state of its outer core. But in fact we do not know for sure.

The magnetosphere is that part of the magnetic field surrounding the Earth that exerts a force strong enough to control the motions of charged subatomic particles entering the field. Exerting a strong force even 50,000 km away from the Earth's surface, the magnetic field protects us from continuous bombardment by cosmic rays; subatomic particles traveling through space at speeds that are a significant fraction of the velocity of light.

From satellites monitoring the magnetic field, we have learned much about the magnetosphere's strength, direction, and composition. Within the magnetosphere are several concentric belts; the principal ones are zones of high particle density known as the Van Allen radiation belts, named after the American physicist James Van Allen (1914-2006) who discovered them in 1958 from Explorer 1 satellite data. The Van Allen belts encircle the planet in two doughnut-shaped regions about 3000 and 17,000 km from Earth's surface.

Charged particles, mainly protons and electrons, populate the magnetosphere's radiation belts. Most of these particles are ejected from the Sun as a reasonably steady flow of matter in the plane of the ecliptic known as the solar wind. When the solar wind particles reach the Earth, they are either diverted away from it or trapped by its magnetic field. The collision of solar wind particles with the Earth's magnetosphere creates a shock wave that distorts and compresses the magnetic field on the sunlit side and stretches it into a long tail on the night side. A shock wave is a compression, such as the sonic boom made by a jet plane.

The trapped particles in the outer Van Allen belt can on occasion spill out and fall into the Earth's atmosphere at high geographic latitudes. There they collide with atoms of oxygen and nitrogen and stimulate these gases to radiate pale greens and occasional bright reds in patches or across the whole sky. These are the auroras, the aurora borealis and aurora australis. They are also known as the northern lights or southern lights, depending upon the hemisphere within which you happen to be. They are most often seen in zones between 65° and 70° north and south magnetic latitudes. Because these particles enter the atmosphere most easily when the solar wind is more intense, more auroras color our night skies during the height of the 11-year sunspot cycle.

6.5. DYNAMICS OF THE MOON

The place of the Moon in human culture, appearing as it does in our literature, music, and art, needs little elaboration. The raising of questions as to its nature, its relationship to Earth, and its origin can be traced all the way back to ancient myths and forward to our own accounts. Although the space age has given us new incite to this nearest cosmic neighbor, many of the old questions are still have not been answered to everyone's general satisfaction. And thus the study of the Moon continues including the desire to return, but this time to establish permanent outposts. Let us begin with a discussion of the dynamical relationship between the Earth and Moon.

The mean distance between the Moon's center and the Earth's center is about 400,000 km. This distance has been measured to within several centimeters by timing the round trip of a laser beam bounced off reflectors left on the Moon by the Apollo astronauts. The Moon's orbit is a small-eccentricity ellipse with the Earth at one focus.

The point that orbits the Sun annually according to Kepler's laws is not the geographic center of the Earth. It is a point on the line joining the Earth and the Moon known as the center of mass. One can think of the center of mass as the center of balance of an imaginary rod supporting the Earth at one end and the Moon at the other. The center of mass for the Earth-Moon system lies inside the Earth, since the mass of the Moon is only about 1 percent that of Earth. Thus in actuality the geometrical center of the Earth orbits the center of mass of the Earth-Moon system with the same sidereal period as that of the Moon, namely 27.3 days. Although such a motion is not readily apparent to us on Earth's surface, it nevertheless has been measured.

The Moon turns once on its axis in the same time that it completes one orbit around the Earth, so that the same hemisphere is always toward us. This is relatively easy to demonstrate to yourself. Walk around a stool, continually facing it; next walk around the stool, keeping your head and body pointed in the same direction. In the first instance, you rotated once while you revolved once, just as the Moon does; in the second, you did not rotate about your axis. If the Moon did not rotate, we could see all its sides during the month. That the Moon's rotation period is equal to the period of its orbital revolution, that is 27.3 days, is not accidental. Tidal forces between the Earth and the Moon over eons have equalized the rotation and revolution periods.

Originally, both bodies were probably much closer, perhaps only 5 to 10 percent of their present distance, and were rotating more rapidly. The Earth's day was a few hours shorter than at present and its month, or the Moon's orbital period, much shorter than now. Because of the Earth's greater tidal force, the Moon's rotation has slowed more rapidly than the Earth's has.

Some of the kinetic energy associated with the Earth's rotation is gradually being transferred by lunar tides to the orbiting

Moon so that the Moon is receding from the Earth by several centimeters every year. The reason for this is that the Moon's tidal force has a braking effect on the Earth, which decreases its quantity of rotational motion, or angular momentum. To conserve the total angular momentum of the Earth-Moon system, the angular momentum in the Moon's orbital motion must be increasing. Hence it is accelerated ever so slightly in its orbit, spiraling outward from Earth.

As the Moon recedes, the month must lengthen, according to Kepler's third law. Eventually, the Earth and the Moon will face each other with equal periods of rotation and revolution, about 47 days, at a distance of about 560,000 km. But the calculated time for this event to happen, several tens of billions of years, far exceeds our estimates of the Earth-Moon system's probable life span.

6.6. THE SURFACE OF THE MOON

The ratio of the mass of the Moon, a satellite, to that of the Earth, a planet, is exceeded only by that of Pluto and its largest satellite Charon. As a consequence of this exceptional situation, we can think of the Earth-Moon system as being more a double planet than a true planet-satellite system. In spite of this, the Moon would not quite span the width of the United States, and its mass is roughly 1 percent that of the Earth.

The program of human exploration of the Moon that began in 1964 with unmanned craft and culminated in six manned Apollo landings between 1969 and 1972 has provided us with a priceless legacy of lunar materials and data. Lunar rocks were collected from nine different locations, six by the United States and three by the Russians, the last being August of 1976. The samples returned amount to more than 2000 individual specimens, weighing about 382 kg (843 pounds).

Five instrument packages were left on the lunar surface, and the last surviving one operated until October of 1977. The seismometers in these packages detected meteoric impacts and many lunar quakes during their operating life span of 8 years.

The Apollo program also carried out an extensive effort to photograph and analyze the lunar surface. The result is maps of some parts of the Moon that are better than those of some areas of the Earth. X-ray and radioactivity studies from orbit have yielded estimates of the chemical composition of about one-quarter of the lunar surface, an area about the size of the United States and Mexico together.

Because of its small mass, the Moon's history has been vastly different from the Earth's. With a small mass comes a weak gravitational attraction; as a result, the Moon retains almost no atmosphere. It has no surface water, either free or chemically combined in the rocks, as in Earth rocks, although some water may be trapped under its surface. It also has no general magnetic field, but its rocks suggest that a strong one existed in the very distant past. However, the Moon is far from a simple, featureless satellite.

Galileo subdivided the lunar surface into maria, the low-lying, roughly circular dark regions, and terrae, the rough, cratered highlands. These are still significant in terms of the lunar history and terrain-shaping processes. The maria are covered with layers of basaltic lava similar to the lava that erupt from terrestrial volcanoes in Iceland, Hawaii, and elsewhere. The highlands consist of a lighter-colored rock that is older than the rocks of the maria. Highlands constitute about 83 percent of the lunar surface, whereas the maria cover only 17 percent.

There are other types of features on the Moon than maria and the cratered terrae. Observers over the years have identified a variety of features, such as a range of sizes of impact craters, rugged mountain ranges, and deep, winding canyons, or rilles.

The lunar mountain ranges tend to be on or near the periphery of the roughly circular maria. The mountains bordering the maria rise more steeply on the side facing them than on the other side. Many have lofty peaks, occasionally rising over 7000 m above the surrounding plains. Although fractures are observed in the lunar crust, there is no evidence that the lunar mountain ranges are folded mountain belts as are those on Earth. Thus, unlike the Earth's mountain ranges, which by and large are the products of collisions between lithospheric plates as discussed earlier, the lunar mountain ranges are the rims of the huge impact basins that contain the maria. Thus their origin is impact cratering and not thermal-tectonic activity as are those on Earth.

Beyond the eastern edge of Mare Imbrium a narrow valley cuts across the Alps Mountains. This feature has long been known from photographs taken from Earth. From photographs taken by an orbiting spacecraft we now know that the Alpine Valley is a deep trough some 3 to 10 km wide and over 100 km long. Narrow channels, known as rilles, which resemble chasms or gorges, cut many kilometers across the lunar terrain, frequently without interruption. Running lengthwise down the middle of the Alpine Valley is a very conspicuous rille, as can be seen in NASA images. Rilles may be lava channels, part or all of which were roofed when filled with flowing lava. Now these tubes have collapsed and are partly choked with rubble from the days of active lava flows.

Over eons of time meteoroids have pulverized the lunar surface, leaving a dusty layer some 1 to 40 m deep that covers the lunar terrain. Known as the regolith, it is the lunar "soil" on which the astronauts left their footprints. Since this soil contains no water or organic matter, it is totally different from soils formed on the Earth by water, wind, and life. More than just bits of ground up lunar rocks, the regolith has also been exposed to cosmic rays, subatomic matter flowing from the Sun, and a fine dust from interplanetary space. Without an atmosphere to shield it, the layers of the regolith contain both the record of lunar events and that of events in the larger Solar System.

A tremendous number of impact craters pit the Moon, evidence of cataclysms that altered the crust during its past. More than 30,000 are visible by telescope. The total, down to bushel basket size, may well exceed a billion. The great walled plains, or supercraters with low profiles, such as Clavius or Grimaldi, have structures similar to those of maria but on a smaller scale. Their diameters are between 200 and 300 km.

Next in size on the Moon's front side are some three dozen impact craters from 80 to 200 km in diameter. A third of them have conspicuous light-colored streaks, called rays, radiating outward in all directions up to several hundred kilometers long, such as the well-known rayed craters Tycho, Copernicus, Kepler, and Aristarchus. Many of the small secondary craters, as well as the ray systems, were apparently formed by a rain of debris ejected from the primary crater after a large body struck the surface.

Impact craters are reasonably circular, with the interior rim steeper than the outer rim. The larger craters have terraces on their inner walls and frequently have a fairly smooth floor from which a few low peaks rise. Beyond the craters the terrain is hummocked and overlain with the ejected material from the cratering process. In those craters having a central peak, the peak is believed to have been created by the elastic rebound of rock from below the surface after the initial impact. Others have bare floors, presumably because they were flooded with lava; the crater Plato is a good example.

The impact craters are not all of the same age. For example, one can see with the crater Clavius that the rim of the major crater is eroded and worn, whereas the half dozen or so small ones in the center have sharper and higher rims. Clearly, the impacting bodies that produced the small craters superimposed on the rim of the large crater must have fallen more recently than that one that formed the large crater. As an example of ages, the craters Copernicus and Tycho are about 600 and 200 million years old, respectively.

Volcanically produced craters may have formed during the Moon's early history. But if so, they are present in considerable fewer numbers than those of impact origin.

Topography on the far side of the Moon appears to the eye to be strikingly different from that on the near side. This is in the sense that impact craters are everywhere, but there are no large lava-flooded basins comparable with Mare Imbrium on the near side, although some small ones do exist. The far side thus lacks the near side's extensive lava flooding. As a consequence, the back side has no extensive mountain ranges.

The Moon's center of mass is displaced from its geometric center by about 2 km earthward. One consequence is that the lunar crust facing the Earth is about half as thick as that of the far side. Perhaps this variation explains why the near side had more volcanic activity in the distant past, which produced the large deposits of dark maria material. It may also help explain why the large impact basins on the far side are only partially filled.

6.7. EVOLUTION OF THE LUNAR SURFACE

The Moon appears to have formed from the same chemical elements, although in somewhat different proportions, as those that formed the Earth. It has less iron, more of the substances that are hard to melt, such as calcium, aluminum, and titanium, and less of the easily vaporized substances, such as sodium and potassium, than does the Earth. The relative abundance of oxygen's three isotopes, however, is extremely close to that of terrestrial rocks. In general, the composition of the Moon mimics that suspected for the Earth's mantle, although it is not a perfect match.

The most common surface rocks are anorthosites, silicates mainly of aluminum and calcium, from the highlands, iron-rich basalt from the maria, and thorium-rich and uranium-rich rocks. No traces of water and no organic compounds, the indicators of living processes, were discovered in any lunar samples. In fact, the Apollo lunar rocks contain only tiny amounts of carbon and carbon-based compounds from which life originates. With no water or oxygen present, the minerals in lunar rocks could not react with water to form clays or rust, nor did iron react with oxygen to form oxides. The lunar highlands, which cover about four-fifths of the lunar surface, are the oldest preserved terrain.

From the findings on the chemical composition of lunar samples coupled with radioactive dates for these rocks, what can be deduced about the history of the lunar surface?

Radioactive dating of lunar rocks points to the formation of the Moon from materials that were much like in composition those forming the Earth. Nearly all the lunar samples are enriched in a kind of mineralogical slag that could only have formed if the Moon underwent a molten phase as did the Earth. This evidence suggests that the Moon probably underwent a global melting to a depth of at least several hundred kilometers, followed by chemical differentiation, a separation of the chemical elements by gravitational settling, shortly after formation. A few of the lunar rock samples are about 4.6 billion years old, the same age as the Earth, even though Earth rocks no longer exist that are older than about 3.8 billion years.

The highland areas are apparently about 4.0 to 4.3 billion years old. After formation of the lunar crust, it was continually modified as a result of impact cratering by material from elsewhere in the Solar System. The cratering record preserved in early crustal units represents a distinct phase of early intense impact cratering, which occurred very early in the history of the Solar System and began to decline about 3.8 billion years ago. Although volcanic processes may have operated during this early period, the surface history of the Moon is primarily that due to impact cratering. As mentioned earlier, this phase in the Earth's history has been almost completely erased.

Impact cratering continues on the Moon today, but at a drastically reduced rate from what it must have been billions of years ago.

The next stage in lunar history was dominated by the formation of dark mare plains, which cover about 17 percent of the lunar surface. These structures are relatively thin ponds of basaltic lava which taken together are less than one percent of the crust's volume. Rocks from maria suggest that the major outpouring of lava occurred between 3.9 and 3.2 billion years ago. Although some mare deposits may be as young as 2 billion years, there appears to have been no extensive lava outpouring on the lunar surface for the last 3 billion years. Thus the shaping of the present lunar terrain is almost the opposite of that of Earth's, the Moon dominated by impact cratering and the Earth by thermal-tectonic activity.

6.8. INTERNAL STRUCTURE OF THE MOON

There is no general lunar magnetic field as large as approximately one ten-thousandth that of the Earth. This seems to indicate that the Moon does not possess a molten iron core comparable with that of the Earth, which is thought to be necessary to produce a magnetic field. But evidence suggests that the Moon may have had a stronger magnetic field early in its history. Random magnetic fields up to about 0.6 percent of the Earth's field intensity were detected at different sites by Apollo astronauts.

From seismographs left on its surface we know that seismic events on the Moon follow patterns different from those here on Earth. Vibrations from moonquakes or rare meteorite impacts are transmitted very slowly through lunar material. They build gradually and then take hours to subside. Some seismic disturbances have been traced to geologic movements in the rilles; others, to occasional impacts of meteoroid swarms. Moonquakes frequently coincide with tidal stresses triggered by the varying distance between the Moon and the Earth. They occur at depths of 600 to 900 km, much deeper than earthquakes. Almost a 100 sources for these deep moonquakes have been discovered so far. But compared with the Earth's seismic activity, the Moon's is fairly subdued; the whole Moon releases less than one ten-billionth of the Earth's earthquake energy.

About 35 shallow quakes, presumably tectonic events, have been detected. Thus in the last 3 billion years, any thermal and geologic activity has been relatively rare. As we have mentioned, most volcanic activity appears to have ceased about 3 billion years ago, but some minor activity may still be going on.

Seismic data tell us that the crust is about 60 km thick, twice the thickness of the Earth's crust. Heat flow from the deep interior through the lunar crust is no more than about a third of that for the Earth. Thus thermally driven processes cannot be nearly as important as for the Earth.

The Moon's mantle, nearly 800 to 1000 km deep under the lunar crust, is uniformly structured. Most of it may be pyroxene and olivine, minerals containing silicon, oxygen, calcium, magnesium, and iron. The seismic data reinforce the view that the Moon's core is unlike the Earth's metallic core. Probably the lunar core consists of partly molten silicates, with a small metallic center. At present, scientists can neither rule out the existence of a small iron core nor prove that one does exist.

Only a few decades ago the Moon was a light in the sky that, even though near, was still part of the remote cosmos. Now it is a place that has been visited by human beings and studied to such an extent that it is no longer a remote cosmic body. What is its future role in human affairs? There are proposals to establish permanent bases on the Moon for astronomical observatories, mining operations, ore refining, or manufacturing; and it is not impossible that the Moon may become the departure station for manned exploration of the Solar System. From any point of view, however, the Moon is no longer a strange and distant world.

Chapter 7
The Terrestrial Planets

Since the Space Age began in the late 1950s, more has been learned about the planets than in all preceding history. By the end of the 20th century we made at least a reasonable beginning to the detailed exploration of all the major planets. Pluto, a dwarf planet, remains to be explored by the New Horizons spacecraft in 2015 if all goes well. Studying different planets, each with its own characteristics, should show us how different planets have evolved to their present state through the same or different sets of dynamic processes, for the planets are the laboratories needed for observing processes beyond the range, in both time and extent, of our terrestrial environment. To understand our own planet better, we need a perspective that can only be acquired by comparative study of the other Terrestrial planets. Having considered the Earth and Moon in the previous chapter, in this chapter we shall relate the progress to date in studies of the other Terrestrial planets. It is not an accident that we have evolved on Earth rather than on Venus or Mars, and we should be aware of that fact when we tamper with our environment.

7.1. THE TERRESTRIAL PLANETS, AN OVERVIEW

Let us begin with a brief overview of the Terrestrial planets. However, since we have just considered the Earth and Moon in the preceding chapter, we will not include them in the overview, but consider the remaining three, that is, Mercury, Venus, and Mars.

Although one of the brighter objects in the heavens, most people have never seen Mercury. It is difficult to study from Earth because Mercury is so close to the Sun. Its maximum angular separation, or greatest elongation, is only 28° on either side of the Sun. At this favorable position for viewing its phase corresponds to a quarter moon; the full phase occurs at superior conjunction, when Mercury is almost in line with the Sun. Swift orbital motion keeps the planet visible low above the horizon for only a few days each year, immediately after sundown or before sunup. From the northern hemisphere, the best time to see Mercury when it is a morning "star" is in the spring, and the best time when it is an evening "star" is in the fall.

Of the Terrestrial planets, Mercury is the least massive, being about 6 percent of Earth's mass, and the smallest in size, being less than twice the size of the Moon. Next to the Earth, however, it has the highest mean density, about 5.4 g/cm^3, a point we shall come back to later.

Mercury's rotation period is two-thirds of its orbital period. That is, the planet completes three rotations during two orbital revolutions. This combination of rotation and revolution periods, like that between the Earth and the Moon, is not accidental; it was apparently set up by the strong tidal force of the Sun, since different parts of the body of Mercury are at slightly different distances from the Sun. Such a tidal force is responsible for having slowed the planet's rotation, trapping it so that the ratio of its rotation to revolution period is 3:2. As a result, from one sunrise to the next one on Mercury takes 176 days; meanwhile the planet completes two orbits of the Sun.

Venus, the second closest planet to the Sun, is the third after the Sun and Moon in brightness in our night sky. Like Mercury, Venus goes through all the phases as does the Moon, as Galileo first observed in 1609. Venus has a larger orbit than Mercury. Thus, Venus swings farther out from the Sun as viewed from Earth, about 47°. Venus remains visible as an evening "star" in the western sky or as a "morning star" in the eastern sky for weeks at a time.

Venus is closest among the Terrestrial planets in size to the Earth with its diameter, mass, and density being only slightly less than those of Earth. Venus possesses a mean density of 5.3 g/cm^3, suggesting that its internal structure is similar to that of Earth and Mercury.

The most striking feature about Venus is a cloud cover that totally hides the surface in visible and infrared radiation. The clouds themselves are almost totally featureless in visible wavelengths, so the planet has a bland appearance with a pale yellow color. In the ultraviolet portion of the spectrum, the clouds do possess features that we will discuss later.

Venus's rotation was a mystery that eluded solution by optical or spectroscopic observations because of its cloud cover and the planet's slow rate of rotation. But Doppler shifts noted in radar observations solved the problem. The planet rotates in a retrograde direction, with its axis of rotation inclined only 2° from the perpendicular to its orbital plane. By retrograde we mean a direction of rotation reversed from that of revolution about the Sun. The period of rotation, as determined from radar measurements, is 243 days. Because its revolution period is about 225 days, just slightly shorter than its rotation period, the Venusian "day" is 117 Earth days long, with 58.5 days of sunlight and 58.5 days of darkness. Thus the Sun rises on the western horizon and sets approximately twice during the Venusian orbit with respect to Earth.

Mars has such an eccentric orbit that the closest approach at opposition between Earth and Mars comes every 15 to 17 years, when Mars is near perihelion in its orbit. At the last most favorable opposition on August 27, 2003, Mars came close enough to Earth that it was about 25 arc seconds wide. At a time such as that even the most casual observers of the heavens are struck by the planet's brilliant ruddy color, far outshining the brightest stars.

Mars is a little more than half Earth's size, has about 11 percent of Earth's mass, and therefore has a mean density 75 percent of that of Earth. The Martian "day" lasts 24 hours and 37.4 minutes. Its axis of rotation tilts 25° from the perpendicular to its orbital plane, giving the red planet seasons like those of Earth, but they last twice as long because the orbital period is nearly 2 years.

Visual observers have made countless detailed maps of the numerous and varied surface features of Mars. Two observers are especially notable: the Italian astronomer Giovanni Schiaparelli (1835-1910) and the American astronomer Percival Lowell (1855-1916), who mapped and named many Martian features. Through a telescope the red planet appears to have Earthlike characteristics, such as white polar caps and large dark areas that vary with Martian seasons.

Recognition that Mars has polar caps

dates back to the early 1800s. When Mars is closest to the Sun, the south pole of Mars is inclined towards the Sun. As a result, the large southern polar cap recedes during the Martian summer, leaving behind a small residual cap or on some occasions none at all that can be seen from Earth. On the other side of the orbit, the north pole is inclined toward the Sun when the planet is farthest from the Sun. The residual polar cap never quite disappears during the Martian summer in the northern hemisphere.

Mars has also for some time been known to have an atmosphere in which vast yellow dust storms occur. Once or twice a Martian year one of these dust storms grows to global proportions, enveloping almost the entire planet in a dense shroud of dust. White clouds in the Martian atmosphere have also been observed from Earth.

Although Giovanni Schiaparelli (1835-1910) and Percival Lowell (1855-1916) gave the public the idea that Mars might have strong similarities to our planet, even life, at least in its past, spacecraft sent to Mars have revealed a surface topography that is perhaps in actuality a bit more like the Moon than the Earth. Landers from Viking (1976 landing) to Curiosity (2012 landing) have offered us a close-up of the Martian surface. There can be little doubt that water once flowed across the surface of Mars in the past. However, the many long-running romantic notions that seasonal changes on the surface of Mars were due to vegetation, and the inference that Mars might thus be the home of intelligent life have long been put to rest.

7.2. TERRAIN-SHAPING PROCESSES

Over the last several decades two events have forced a revolution in our thinking about planets, particularly the surfaces of the Terrestrial planets. One is the recognition of the Earth's thermal-tectonic activity and the other is the combined results from planetary space exploration by the United States and Russia.

Almost everywhere across the surfaces of the Moon, Mercury, and Mars we see evidence of impact cratering and volcanic activity, with the relative percentage of the two varying from one body to the next. In addition, from radar studies of Venus it appears that evidence is there of both processes. For Earth, however, scientists have a very different picture: It is a planet presently dominated by thermal-tectonic activity.

During the first billion or so years of the Solar System a period of heavy bombardment, called the intense impact-cratering period, produced most of the impact craters seen on the Terrestrial planets. A byproduct of the intense impact cratering was flooding with lava of immense impact basins to form the maria. This intense cratering period explains essentially why the surfaces of the Moon and Mercury look as they do. However, for the larger Terrestrial planets, that is Earth, Venus, and Mars, the second major terrain-shaping mechanism has remade almost all the surface in the case of Earth and part of the surfaces in the cases of Venus and Mars. This mechanism is the thermal-tectonic activity produced by convection in the mantle.

The surfaces of Venus, Earth, and Mars over billions of years were fractured, deformed, and in the case of Earth worn down by erosion from water, wind, and life. Volcanic material can reach a planet's surface to develop plains or volcanoes only if the crust can be broken, allowing lava to flow onto the surface. Over the 4.6-billion year existence of the Terrestrial planets, the Earth has shown by far the greatest volcanic activity, with presumably Venus next, followed by Mars, Mercury, and the Moon.

If heavily cratered terrain is the oldest type of surface, then its preservation implies a relatively stable history for a crust. Why should the smaller Terrestrial planets have old and, consequently, relatively stable crusts, whereas the larger Terrestrial planets have younger, in some cases much younger, portions to their surfaces because of thermal-tectonic activity? Just after the formation of the Terrestrial planets, these planets must have been completely molten spheres of rocky materials. Within a few hundred million years after formation, their surfaces solidified forming a thin crust that enclosed a molten interior. Large bodies bombarding their surfaces, however, could still puncture or crack the crust allowing molten material from the interior to flow onto the surface in great seas of lava.

The important point is that the Terrestrial planets would not have cooled at the same rate. This is because the thermal energy content of a planet depends on its volume, which is proportional to the cube of its radius. Whereas the rate that a planet cools will depend on its surface area, through which thermal energy is lost to space, and the surface area is proportional to the square of the planet's radius. Thus the ratio of a planet's thermal energy content to its rate of cooling is proportional to its radius, or radius cubed divided by radius squared, meaning that smaller planets radiate away energy more rapidly than do large planets. Smaller planets, consequently, cool faster than do larger planets.

Therefore, the Moon, being the smallest of the five bodies, should have cooled the most rapidly forming a thick protective crust which stopped the surface from undergoing significant alterations at least 3 billion years ago. The next to form a protective crust and cease radical surface changes should have been Mercury, since it is only slightly larger than the Moon, followed by Mars. Venus is still undergoing some major changes in its surface terrain. However, the Earth's surface is still very much "alive," unlike the "dead" surface of the Moon, and should continue to experience dramatic change for the next 2 billion or so years.

7.3. SURFACES OF THE TERRESTRIAL PLANETS

Observations from the Earth had hinted that Mercury might look like the Moon. But it was the three Mariner 10 flybys between March of 1974 and March of 1975, and the MESSENGER (MErcury Surface, Space ENvironment, GEochemistry, and Ranging) spacecraft, between 2008 and 2013, that have showed that the planet is indeed heavily cratered like the Moon. Although the overall surface of Mercury is remarkably similar to that of the Moon, there are significant differences. These differences suggest a somewhat different surface evolution from that of the other Terrestrial planets.

The surface of Mercury is pockmarked

with craters ranging from hundreds of meters up to hundreds of kilometers across. Some of the bright craters have extensive ray systems like those on the Moon. Compared to the Moon, Mercury and Mars are deficient in craters in the range of a few tens of kilometers. There are some conspicuous differences between Mercury and the Moon, however. Craters on the Moon's highlands are densely packed, with rims of young craters overlying old craters, and the mare regions possess sharp boundaries. On Mercury, by contrast, craters are often interspersed with relatively smooth plains, giving the terrain a speckled appearance.

Like the Moon, both Mercury and Mars have dark maria. The maria for Mars and the Moon are almost indistinguishable; however, maria on Mercury exhibit small but significant differences. Twenty or so of Mercury's maria are several hundred kilometers across, while Caloris, the largest one is more than 1000 km wide. The interior surface of Caloris Basin resembles Orientale Basin on the Moon. Scarps or cliffs a few kilometers high and often hundreds of kilometers long cut across maria and craters alike.

The basins of the maria and many other features on Mercury, as on the Moon, were created in the first half billion or so years of its existence during the intense impact-cratering period. Because Mercury and the Moon cooled at a faster rate than did Venus, Earth, and Mars, the results of this intense cratering period can still be seen on the surfaces of Mercury and the Moon, and they have remained virtually unchanged for at least the last 3 billion years.

Venus's surface is hidden by total cloud cover. Radar studies suggest that the planet's surface possesses geologic features suggestive of impact cratering, volcanism, and primitive tectonic activity. Pioneer Venus and Magellan radar studies from orbit confirmed Earth-based observations. While 65 percent of the Earth's surface is ocean basin lying on average about 5 km below sea level, fully 60 percent of Venus's surface lies within 0.5 km of the mean radius of the planet, and only 5 percent is more than 2 km above the mean radius. Despite this limited spread in elevation, the maximum distance between the highest and lowest points is about 13 km, which is comparable to that on Earth.

There are two continent-like features rising well above the average surface level. One highland area, Ishtar Terra, is unlike anything seen on the Moon, Mercury, or Mars. It is larger than the continental United States and stands several kilometers above the mean planet radius. In elevation it is similar to the Tibetan Plateau on Earth but about twice its size. Maxwell Montes on the eastern end of Ishtar Terra contains the highest point on Venus's surface, about 11 km above the mean radius or 2 km higher than Mount Everest is above sea level. The Venusian mountains containing Maxwell Montes rise out of the Ishtar Terra plateau in the same way that the Himalayas stand on the Asian plate. An artist's rendering of this continent-like feature is available from NASA. The other continent-like region, Aphrodite Terra, is also comparable in size with continental United States.

Also disclosed by radar observations of Venus's hidden surface are some almost circular features between a few tens of and a thousand kilometers in diameter which may be impact craters and basins. From Venera 15 and 16 and Magellan radar images that provide better resolution of the surface, it appears that some of these circular features are volcanic caldera flanked by large lava flows. Other large features include a 1000 km long trough a few hundred kilometers wide and a few kilometers deep which is similar in scale to Valles Marineris on Mars. Maxwell Montes is a large, low, circular dome some hundreds of kilometers across with a central depression 100 km or so in diameter, similar in many respects to a volcanic peak. If truly a volcano, it is about 25 percent larger than Olympus Mons on Mars. There are two other regions of suspected volcanic activity. These are Beta Regio and the eastern end of Aphrodite Terra.

Although there has been no direct observation of a volcanic eruption, several lines of evidence suggest that major volcanic activity has occurred within the last few tens of millions of years, and according to some studies maybe as recently as the last 50 years. This evidence includes the Magellan radar images of possible volcanic craters and lava flows, major changes in the amount of volcanic gases in the clouds, and radio noise characteristic of lightning around volcanic peaks. Not one of these arguments by itself is entirely convincing, but taken together they present very persuasive evidence for active volcanoes on Venus. Since there is no clear evidence for lithospheric plate formation and subsequent motion, Venusian volcanoes are probably isolated hot spots in the crust like Mauna Loa in Hawaii rather than like the strings of steep volcanic cones that characterize Earth's subduction zones where one plate is being forced under another.

The two Russian landers, Venera 9 and 10, sent back the first photographs of the Venusian surface in 1975. Venera 13 and 14 provided four more, with at least one in color. Sunlight filtering through the cloudy atmosphere supplies enough light to make the surface look like a dark, overcast day on Earth. The atmospheric color is decidedly orange-like, since blue photons of incoming sunlight are absorbed and scattered by clouds. Venera 13 showed a rock-strewn plain with a dark, fine-grained material interspersed between rock outcroppings. Although having a number of noticeable differences, the view roughly resembles Martian terrain. Venera 14 landed on a terrain different from that for Venera 13. Its view was of a plain of broken rock layers that extend to the horizon. Samples were collected by both spacecraft and analyzed. Their composition suggests that the material is basaltic rock, an igneous rock extruded from the interior and generally silicon-poor and metal-rich. Such basaltic rock is common on the Earth and the Moon.

The impact craters and basins on Venus suggests a surface billions of years old, while the possibility of volcanoes, rift valleys, and plateaus point to youthful parts of the surface that may be only millions of years old. Unlike Mercury and the Moon, Venus appears to be a relatively active planet that may resemble the early Earth.

With respect to Mars, the fine, delicate streaks called "canals" sketched by observers on early Martian maps are quite clearly illusory. Missions to Mars including the early Mariner and Viking pictures revealed these canals to be dark-floored craters or irregular dark patches aligned by chance and linked unconsciously by early observers into lines that looked like canals.

Like the Moon and Mercury, Mars has a different topographic pattern in each hemisphere, but it is more diverse and complex than either the Moon or Mercury. Mars's northern hemisphere is generally lower than the mean radius of the planet by a couple of kilometers, possesses few craters, and has been altered by intense volcanic activity. The extensive lava flooding occurred at various times after the cessation of the intense impact-cratering period. The southern hemisphere, however, has a densely cratered surface, averaging a couple of kilometers greater than the mean radius. Its crust has not changed appreciably throughout the planet's life.

There are about 16 smooth, circular basins containing lava-flood plains on Mars. One that has long been observed from Earth is Hellas, an almost craterless basin about 1800 km wide or about one and a half times the size of the largest lunar sea Mare Imbrium. The small number of impact craters on the maria suggests that they formed after the intense impact-cratering period ceased some 4 billion years ago. Estimates of the age of the ancient lava plains are about 3 billion or so years.

The abundance of craters in some regions of the southern hemisphere is comparable with that in the highlands of the Moon. The similarities, between the cratered Martian southern hemisphere and the lunar highlands, prompt speculation that the two are about the same age. Thus almost half the surface of Mars is ancient terrain, with many of its landforms having remained essentially unchanged over the last 4 billion years.

From spacecraft photographs one sees that Mars has a wide variety of channels in the oldest terrain. They appear to have formed between 3 and 4 billion years ago, shortly after the end of the intense cratering period. We have photographs of channels that have the appearance of some drainage systems on Earth. Such images were the first evidence that water flowed on the Martian surface in the past and is now held as subsurface ice. It is apparent that there was a time in the Martian past when a thicker and warmer atmosphere permitted liquid water to exist on the surface.

An important departure from the character of the major portion of the Martian terrain is the Tharsis ridge. This area is different because of three large volcanoes running diagonally along the crest of the ridge and a spectacular, isolated volcanic structure, Olympus Mons, which is similar to, but much larger than, Mauna Loa and Mauna Kea in Hawaii as seen from the bottom of the Pacific Ocean. In addition to these large shield volcanoes rising some 20 km above the surrounding plains, there are flattish saucer-shaped volcanoes, some of which lack a significant number of impact craters on their slopes, suggesting that they are relatively young. Tharsis ridge is 10 km or so above the average Martian radius; the volcanoes extend above it. Like the suspected volcanoes on Venus, Martian volcanoes are more akin to Earth's "hot-spot" volcanoes in the middle of a plate, such as Mauna Loa in Hawaii, rather than to lines of volcanoes setting astride plate boundaries where one plate is forced underneath another.

In the equatorial region lies the spectacular canyon Valles Marineris, which cuts across the middle of a plateau. It is nearly 4000 km long, up to 100 km wide in some places, and at least 4 km deep. At its western end lies a complex pattern of intersecting fault valleys. Valles Marineris runs radially away from Tharsis ridge and probably results from the faulting that accompanied the evolution of Tharsis since its formation. However, thermal-tectonic activity on Mars is much weaker than that on Earth.

From orbit, the dominant features of the region around the Viking 1 Lander are craters. From the ground, there are only a few obvious craters in the immediate vicinity of the Lander. If Mars were like the Moon, then there should be visible several small craters that are tens of meters in diameter. Their absence indicates that the Martian atmosphere is dense enough to burn up small meteoroids or material ejected from large impacts before they reach the surface and that surface erosion is vigorous enough to obliterate small craters. Thus there is not the profusion of small craters seen on the

Moon.

The area photographed by the Viking 1 Lander in the Chryse region is a gently rolling landscape, yellowish brown in color, strewn with rocks and dotted with drifts of fine-grained material. Within 30 meters or so of the Viking 1 Lander lie several outcrops of bedrock, which in many ways resemble the semi-desert regions of the American Southwest but without vegetation. Chemical analysis by both Viking Landers and the later rover missions, including Curiosity, found elements such as silicon, oxygen, iron, magnesium, and aluminum, or those common in Earth's crust. However, organic analyses by the Viking Landers and the Curiosity Rover failed to detect any organic molecules, that is, those molecules that contain carbon, one of the essential ingredients for life. On Earth, soil in even the most dry, sterile-looking valleys contains many organic compounds.

The multilayered polar regions are still another type of Martian topography. The layered deposits hold appreciable quantities of frozen water mixed with dust beneath a carbon dioxide coating, as verified by the

Phoenix Lander. Periodic changes in the Martian climate may be responsible for the deposition of the successive layers of material.

As we discussed in the previous chapter, evidence from the intense impact-cratering period has been erased from the Earth's surface. Even the oldest rocks, which are 3.5 to 3.8 billion years old, are of no help in identifying this early intense cratering period because they are mostly isolated outcroppings, many covered with ice. However, later impact craters still exist. Earth also has basaltic plains, like maria on the Moon, Mercury, and Mars; not old ones, as on these Terrestrial planets, but very young ones. The major ones on Earth are formed by the addition of new material to the lithospheric plates at the mid-oceanic ridges.

It is unlikely that Earth ever went through an appearance like that of the Moon, Mercury, and Mars, in which maria were formed from lava's flooding a huge impact basin. In all probability, the thermal-tectonic activity of Earth's surface has always been too great to have allowed maria to last any appreciable length of time. Thus if we could watch a time-lapse movie of the evolution of the surface of all the Terrestrial planets, it is unlikely that they would all start out the same and begin to depart from each other later. Rather, the surfaces of the Moon and Mercury, which have thick lithospheres, thicker than those of Mars, Venus, and Earth, have not been fractured and then deformed by convection. This is so because the Moon and Mercury cooled very quickly, extinguishing any tectonic activity if it ever existed. And, the Earth has probably always shown vigorous tectonic activity, with Venus's less than that of Earth and Mars even less. Thus the Moon and Mercury have the oldest surfaces, Mars's surface is old but with some youngish features, Venus one guesses to have a mixture of old and young, and Earth's surface is comparatively very young.

7.4. PLANETARY ATMOSPHERES

We are all aware of the characteristics of Earth's atmosphere and its consequent importance in maintaining life; Mercury and the Moon, in contrast, have almost no atmosphere, while Venus and Mars have carbon dioxide atmospheres and the Jovian planets have extensive hydrogen-helium atmospheres. How did this diversity in planetary atmospheres arise?

There are several factors that determine why a planet's atmosphere is the way it is, this includes:

- the planet's distance from the Sun along with its size and mass determine what its chemical composition and temperature will be, which in turn influence its ability to retain an atmosphere;
- its chemical composition along with its temperature determines what chemical processes go on in the atmosphere;
- geologic and chemical evolution of the planet's surface layers influence the evolution of the atmosphere;
- and, if living organisms are present on the planet, they will influence the nature of the atmosphere through long-term interaction with it.

We will consider each of these factors in

our discussions in this and the next chapter.

A planet's distance from the Sun and its mass are important in determining which molecular constituents it can retain in its atmosphere. From our earlier discussion on random thermal motion, we know that the higher the temperature or smaller the mass of a molecule, or both, the greater will be the average thermal velocity and the more likely that gas particles can escape an atmosphere. Using a planet's temperature, mass, and radius and the masses and thermal velocities of its atmospheric constituents, astronomers conclude that if a molecular constituent's thermal velocity is near one-third the escape velocity, about half that chemical species will escape from the atmosphere within weeks. For a planet to retain various molecular components indefinitely, the mean thermal velocity must be less than a tenth of the escape velocity.

The massive Jovian planets, with their large escape velocities, several tens of kilometers per second, have held their primeval atmospheres of hydrogen and helium, whereas the less massive Terrestrial planets, with smaller escape velocities, several kilometers per second, have lost, if they ever had, these light gases. Venus, Earth, and Mars, however, have managed to retain atmospheric water molecules as well as such heavier gases as carbon dioxide and nitrogen. Mercury and the Moon lack any appreciable atmosphere because these two bodies have small masses; moreover, Mercury being quite close to the Sun is consequently very hot.

Chemical behavior is an important factor affecting a planet's atmosphere, which for Terrestrial planets is very different from that of Jovian planets. Because the Terrestrials formed from rocky materials, their early atmospheres should have been largely composed of such gases as carbon dioxide, nitrogen, and some water. Venus and Mars still have that kind of atmosphere. However, planets such as Jupiter and Saturn are more nearly like the Sun in chemical composition than like the Terrestrials and thus show chemical behavior different from that of Terrestrial planets.

The incoming rays from the Sun, which are primarily in the visible part of the spectrum, penetrate a planet's atmosphere of clear gases and are absorbed by the surface layers, warming them. In turn, the surface reradiates energy in the infrared region of the spectrum back out into the coldness of interplanetary space. The reradiated energy is in the infrared region because the warming of the surface by sunlight maintains it at a temperature of only several hundred Kelvins. Passage of the reradiated infrared photons outward through the atmosphere into space, however, is hindered by any infrared absorbers such as carbon dioxide and water vapor. They absorb infrared photons and in turn reradiate much of the energy back to the surface. This process is called the greenhouse effect, after the similarity of the process to that of the glass in a greenhouse, which prevents heat radiation produced inside a greenhouse from escaping.

Depending on the chemical composition and chemical activity of a planet's atmosphere, the greenhouse effect will be more or less effective in trapping incoming solar radiant energy. This causes a warming of the low atmosphere and surface, which can profoundly affect the atmospheric chemistry. Venus is a good example of the long-term consequences of the greenhouse effect. Estimates are that the mean surface temperature of Mars, Earth, and Venus are about 5K, 35K, and 500K warmer, respectively, than they would be without the greenhouse effect.

7.5. ATMOSPHERIC STRUCTURE AND EVOLUTION

Mercury's atmosphere is very tenuous, approximately a million billion times less dense than that of Earth. It seems to be supplied and constantly replenished by the solar wind. Helium and a couple of percent of atomic hydrogen have been identified as its principal constituents. The abundance of other atomic or molecular species, if they are present, is insignificant. No evidence has been found for atmospheric modification of any landform.

Some 4.6 billion years ago, just after Mercury formed, gases such as carbon dioxide escaping from the planet's interior may have temporarily created an atmosphere of some extent. But soon after forming, it would have escaped into space and vanished. This is because the planet is not massive enough to hold much of an atmosphere at such a small distance from the Sun and probably was endowed with less gaseous material during its formation than other Terrestrial planets.

The atmospheric pressure on Venus's surface is about 90 times greater than that of Earth while the surface temperature is some 2.5 times greater. Analysis of the lower atmosphere suggests that it is about 96 percent carbon dioxide and 3.4 percent nitrogen, with the remainder water vapor and some other gases. Because of the high surface temperature on Venus, carbon dioxide was apparently not depleted, as it was on Earth, by reacting with primitive rocks to form carbonates and limestones and by absorption by water. Above 150 km atomic oxygen is the most abundant species. And finally, a huge cloud of hydrogen surrounds the planet far above the atmosphere.

More speculation than on almost any other aspect of Venus has been given to the mysterious clouds that perpetually obscure the surface. In 1973, it was suggested that the clouds were composed of sulfuric acid droplets. From Pioneer Venus results, it appears that the clouds are indeed composed of sulfuric acid droplets and other particles, possibly free sulfur, so thick that during the descent of the probes they appeared to be passing through a blizzard. Data suggest that several other sulfur compounds also exist in the atmosphere. Although once thought to be an unchanging aspect of Venus, the clouds are probably continuously produced and destroyed rather than being an unchanging feature of the atmosphere. A key link in the chemical cycle producing the clouds is volcanic activity on the surface. The extent of the sulfuric acid clouds is controlled by a complicated cycle of atmospheric and crustal chemical reactions.

The clouds begin around 46 km above the planet's surface and seem to be confined to a fairly distinct layer, rising up to about 70 km. A thin haze exists above and below the cloud layer; the lower one has a surprisingly abrupt cutoff some 32 km above the surface. From the bottom of the haze down to the surface the atmosphere appears to be surprisingly clear.

The atmospheric circulation is the same in both hemispheres. A vigorous equatorial east-west jet stream is quite evident in the upper atmosphere, moving around the planet in only 4 days, opposite to the direction of the planet's slow spin. The wind velocity decreases at lower altitudes until at the surface it slows to a gentle breeze. The lower atmosphere apparently circulates because of differences in solar heating between the equatorial and polar regions. Clouds rise near the equator, spiral toward the poles, and descend in what appears to be an almost continuous flow. But we do not know why such high winds reverse direction in the upper atmosphere.

As on Venus, carbon dioxide is the most abundant constituent in the thin Martian atmosphere, amounting to about 95 percent. We know that the atmosphere also contains about 2.7 percent nitrogen, about 2 percent argon, lesser amounts of atomic and molecular oxygen, and traces of other constituents.

The warmest daytime temperature is around 30° C at the Martian equator, while the nighttime temperature drops to -130° C. Over the polar regions it is even colder. During summer the north polar ice cap gets up to only -70° C, while very cold, not cold enough for the residual cap to be made of carbon dioxide ice. Thus it appears that it is water ice, which is consistent with finding more water vapor in the atmosphere at high latitudes near the poles.

A small, daily, and seasonally variable amount of water vapor has been monitored in the atmosphere. However, because of the low atmospheric temperature and pressure, less than 1 percent of the Earth's sea level atmospheric pressure, near the surface of Mars, the amount of water vapor is far too low for rain or for water to exist as a liquid on open, flat ground.

Early morning fog lying in craters and other low places is probably evidence of an exchange of water vapor between subsurface or surface ice and the atmosphere. The Martian atmosphere also possesses clouds, which are most probably condensations of water and carbon dioxide ice.

One of the most exciting events during the active life of the Viking Landers was the photographing of frost on the surface of Mars at the Viking 2 site. The frost occurred during the northern winter, between May and November of 1977. The composition of the frost is not known; the atmospheric temperature was too warm for it to have been pure carbon dioxide ice, and the atmosphere was too dry for it to have been pure water ice. The best speculation is that it was a mixture of carbon dioxide and water.

The skies at the locations of Viking 1 and 2 were yellowish brown in color and seemed to remain that way over the course of the Martian year. This color was probably due to dust particles suspended in the atmosphere below about 50 km. Surface winds can stir the atmosphere sufficiently to hold dust particles. From Viking data we know that the prevailing winds are westerly, as on Earth, with velocities up to 70 km/h at the surface and over 360 km/h at altitudes above 10 km. In fact, the winds are strong enough to create major dust storms that can engulf almost the whole planet and last for months. Undoubtedly the winds come from unequal solar heating of the Martian surface, driving air from high to low pressure areas, as on Earth.

Since the Sun contains significant noble gas abundances, the planets, having formed from the same basic material, should possess a significant amount of these heavy gases in their atmospheres. These gases are particularly useful in evolutionary studies because they are too heavy to readily escape into space, and they are chemically nonreactive and, consequently, are very difficult to incorporate into surface rocks. The surprising scarcity of neon, argon, krypton, and xenon in the Terrestrial planets' atmospheres suggests that atmospheres on Venus, Earth, and Mars probably formed from gases escaping from their interiors during volcanic eruptions.

Volcanic outgassing from the interiors of Venus, Earth, and Mars early on should have consisted primarily of carbon dioxide, nitrogen, and water vapor in approximately the same proportions that we observe in Terrestrial volcanic gases today. On Earth, water vapor condensed to form oceans, but nitrogen remained in a gaseous state. Most of the carbon dioxide combined with silicate rocks in the crust to form carbonate rocks, such as limestone, a reaction that occurs most efficiently in the presence of liquid water. If carbon dioxide could be somehow released from crustal rocks along with the small amount dissolved in the oceans, the amount in the Earth's atmosphere would equal about one-half that of the dense atmosphere of Venus.

If one assumes that Venus formed with about the same relative amount of water as Earth did, then the challenging question is what happened to it, since it is not on the surface in pools nor in the atmosphere. Venus receives about twice as much radiant energy from the Sun as does Earth, so that its atmospheric temperature should have been higher from the beginning. Because of the high temperature, water most likely stayed in vapor form, which would also have been the case for carbon dioxide. Eventually, these two molecules could have produced a runaway greenhouse effect that amplified the evaporation of water. Once the water vapor was part of the atmosphere, incoming ultraviolet photons from the Sun could dissociate the water molecules, allowing free hydrogen to escape over the planet's lifetime and the heavier oxygen to combine with crustal rocks to form oxides. Some water is also consumed in making sulfuric acid droplets in the clouds, and a tiny amount of water vapor is still present in the atmosphere.

On Mars, the outgassing of water vapor, carbon dioxide, and nitrogen was probably less complete than it was on Earth, yet carbon dioxide forms the largest part of the atmosphere of Mars. Mars appears at present to be in a cold phase, and a large amount of water is apparently stored in the polar caps and under the planet's surface as permafrost. There is speculation that such water ice may be a remnant of a denser atmosphere that Mars had in the first billion years or so of its existence. Even though Mars has always received less energy from the Sun compared to Earth, if that early atmosphere was a denser carbon dioxide, say 100 to 200 times more than at present, and water vapor atmosphere, then it could have acted to trap infrared radiation through the greenhouse effect. And this could have made the atmosphere warm enough to contain substantial amounts of water vapor. This increased amount of carbon dioxide could easily have been provided by outgassing from the body of the planet. However, over time, the formation of carbonate rocks removed carbon dioxide from the atmosphere and lowered both the temperature and pressure. Under such conditions, the atmosphere could no longer retain much water, and water could not exist as a liquid on the surface. Thus the era of water ended for Mars some time ago, with death due to the cold rather than heat, as was the case with Venus.

Why did the escape of water and consequent heat death on Venus or freezing of water and consequent cold death on Mars not occur on Earth, since its early atmosphere was probably similar in composition to those of both Venus and Mars? It is probable that the advent of photosynthesizing life on Earth began to replace carbon dioxide with oxygen about 2 billion years ago and prevented a substantial greenhouse effect; thus water stayed primarily in pools on the surface. Large expanses of liquid water moderated Earth's climate and provided an environment conducive to the further development of life, given there was adequate protection from solar ultraviolet radiation. Most of Earth's free oxygen, so necessary to animal life, comes from photosynthesis, through which oxygen is constantly replenished by green plants, plankton, and some bacteria. When living organisms began to extract carbon dioxide from the atmosphere, they helped save the Earth from the heat death that Venus has apparently experienced.

It has been known for many years that Earth has a magnetic field that extends far beyond the body of the Earth to form a magnetic envelope, or magnetosphere. Earth's magnetic field is attributed to electric currents flowing in the outer core, so the Earth is somewhat like a giant electromagnet. If other Terrestrial planets also have electric currents flowing in their deep interiors, then they too will have magnetospheres. Or we can turn the argument around: Detection by spacecraft of a magnetic field tells us something about the interior of a planet, that is, whether or not it is sufficiently fluid for matter to flow in the deep interior to generate electric currents.

An unexpected discovery of the Mariner 10 mission was that Mercury has a shock front similar to the wave surrounding the bow of a ship plowing through water. Like the one for Earth, Mercury's bow shock is caused by the onrushing solar wind particles colliding with the planet's magnetic field. Although Mercury does have a field, it is only 1 percent as strong as that of Earth. Consequently, it is much too weak to hold radiation belts such as Earth's Van Allen belts. The magnetic axis of its field almost coincides with Mercury's axis of rotation.

Before the advent of planetary exploration with spacecraft, astronomers suspected that Earth's sister planet, Venus, might have a magnetic field comparable with and produced in the same way as Earth's field. However, after data from the first few Mariner and Venera missions were in, it was evident that the Venusian magnetic field is much smaller than Earth's field. Venus has a well-developed bow shock formed by solar wind particles impinging on the planet's magnetic field; but that weak field, like that of Mercury, is too feeble to trap solar wind particles in radiation belts.

The weakness of Venus's magnetic field is probably due to the planet's very slow rotation.

Mars's magnetic field intensity is much less than that of Earth. There is apparently a feeble bow shock formed between the onrushing solar wind and the magnetosphere, but no radiation belts appear to exist.

In summary, Earth is the only Terrestrial planet with a field intense enough to retain radiation belts. This suggests that Earth is the only Terrestrials possessing sufficient circulation in a fluid core to generate a strong magnetic field.

7.6. INTERNAL STRUCTURE OF THE TERRESTRIAL PLANETS

The final aspect of the Terrestrials that we want to consider is their interiors; those regions hidden from direct observation. We overcome this limitation by calculating mathematical models of the interior from a planet's observed physical properties and from theoretical arguments about the physical laws governing a planet's internal structure. Ideally the physical properties needed are a planet's mass, mean density, shape, rotation rate, gravitational and magnetic field strengths, surface temperature, and chemical composition. However, even with less than all these pieces of information, an interior model can be calculated, but it is accordingly more speculative. Naturally, interior models would be more accurate if we had seismic data and rock samples, as we have for the Earth and Moon. Planetary interior models, and as we shall see later interior models of stars, are two of the prime examples of the use of scientific models to guide our thinking in the pursuit of knowledge.

When constructing an interior model for a planet, astronomers make the following assumptions about the appropriate physical processes going on inside planets:

- Since a planet has a stable configuration, that is it is neither contracting nor expanding, the weight of matter caused by gravity pressing inward is balanced by the pressure of matter deeper inside pushing out.
- The pressure exerted by interior

matter depends on its density and temperature in a more complicated fashion than is true of a simple gas.

• The flow of heat out of the deep interior determines what the decrease in temperature will be out through the body of the planet.

• Finally, interior matter can change from solid to liquid, can deform and flow under pressure, and can form different types of mineral compounds.

Having constructed a model, astronomers can use it to predict how temperature, pressure, and density vary from the planet's center to its surface. Planetary models essentially show that other Terrestrial planets, like Earth, possess such layered zones as core, mantle, and crust, whereas Jovian planets have a different type of structure.

The values of its mass and radius imply that Mercury must contain a large fraction of iron, the only heavy element sufficiently abundant to account for the planet's high mean density. The iron and nickel content may be as much as 65 percent of Mercury's mass. By analogy with terrestrial, lunar, and meteoritic chemical abundances, we presume that silicates and oxides of iron are also prevalent. Additional evidence for a large iron-rich core comes from Mercury's magnetic field, which is intrinsic to the planet and is most likely the result of an internal mechanism that continuously generates the field in much the same way as does Earth. Chemical differentiation appears to have occurred very early in the planet's history, probably during the first half billion years. Since then, the surface has been largely undisturbed by thermal-tectonic processes.

The model that has been derived for Mercury is one with a crust overlying a silicate mantle, which in turn surrounds a molten, or partially molten, iron-rich core. The core radius may be as much as 76 percent of the planetary radius, a percentage that is considerably greater than that for any other Terrestrial, including Earth. Such a core should be adequate to generate the magnetic field observed by Mariner 10 and MESSENGER.

Since Venus is the Terrestrial planet closest to Earth in size and mass, it is reasonable to expect that it will generally be the planet most like the Earth. Of critical importance in developing a model for its interior is Venus's chemical composition: As we have seen from estimates of what chemical elements were likely to have been present at the time of formation of the Terrestrial planets and from observations of the planets' influences on the motion of spacecrafts, astronomers have made estimates for the iron and nickel content in the Terrestrial planets; as percentages of total mass, the iron and nickel content is as high as 65 percent for Mercury, up to 38 percent for Venus, up to 33 percent for the Earth and Moon, and only up to 26 percent for Mars.

Such an iron content for Venus suggests that it ought to have a molten core, like that of Earth, that occupies about the same fraction of Venus's interior as does Earth's, or significantly smaller than the core of Mercury. Overlying the molten iron-rich core is a silicate mantle, and there is a crust on top of the mantle. It is possible that Venus, also like Earth, has an inner core of solid iron-rich material. Speculation for and against an inner core depends on estimates of the iron content of the planet.

Besides the Earth and Moon, Mars is the only planet for which scientists have seismic data. Both Viking Landers carried instruments to record quakes on Mars. Unfortunately, only the one on Viking 2 worked, and in November of 1976, it is believed by some astronomers to have detected a quake. Any seismic activity on Mars, although it should be somewhat more extensive than that of the Moon, should also be much less than that of Earth.

If valid, the Mars quake seismic data suggest that the average thickness of Mars's crust is greater than that of Earth. Earth's crust is about 0.5 percent of its radius, Mars should be closer to 1 percent, whereas that of the Moon's crust is about 4 percent. This suggests for Mars a lithosphere, that is crust and outer portion of the mantle, a couple of hundred kilometers thick on a body, which is presumably chemically differentiated. If so, Mars should have a silicate mantle and an iron-rich core of about 1500 km radius. Thus, of the Terrestrial planets, Mars has the smallest percentage of iron and the smallest core for its size.

This completes our survey of the Terrestrial planets. We now move in the next chapter to a survey of the Jovian planets, which are the principle occupants of the outer Solar System.

Chapter 8
The Jovian Planets

Our knowledge of the inner part of the Solar System is considerably greater than that of the outer part: There are few or no data on many aspects of the outer parts of the system. And while the boundaries of the inner Solar System are reasonably well defined, those of the outer are not. For example, is it possible that there are small, faint, distant planets beyond Pluto, awaiting discovery? The discussion of the planets previously, demonstrated that there are distinct differences between the Terrestrial and Jovian planets. For example, the four giant planets, Jupiter, Saturn, Uranus, and Neptune, contain 99.6 percent of the total mass of the Sun's planets. And Jupiter and Saturn, with their large complements of satellites, are in some regards like miniature Solar Systems. Certainly the differences in the compositions of these bodies compared with the compositions of the Terrestrial planets suggest that the differences between Terrestrial and Jovian planet have existed since the Solar System began.

8.1. THE JOVIAN PLANETS, AN OVERVIEW

As we did earlier for the terrestrial planets, let us begin our discussion of the Jovian planets with an overview of the general features of these bodies. After which we will go into a comparative discussion of more specific aspects.

Fifth planet from the Sun, Jupiter is the largest and most massive of the planets in the Solar System. In our night sky it glows with a bright, steady yellow light, outshining the stars. The mean diameter of Jupiter is about 11 times greater than that of Earth and Jupiter is more than 1000 times larger in volume than Earth. Jupiter's mass, however, is barely more than 300 times that of Earth, even though it exceeds the combined masses of all the other bodies orbiting the Sun. Thus its mean density is about one-fourth that of the Earth. Because its axis is tilted only 3° from the perpendicular to its orbital plane, the planet has little seasonal change.

Not all portions of the visible layers of Jupiter, which appear as alternating dark and light bands parallel to the equator, rotate in unison. The equatorial region completes its rotation several minutes sooner than adjacent higher latitudes. This phenomenon is known as differential rotation and is possible in fluid media, such as gases. It is not something one expects a solid body, such as the surface of a Terrestrial planet, to do. Jupiter's rapid 10-hour rotation and low density combine to flatten the planet about 6 percent in its polar diameter. Again, this is more characteristic of a fluid body that will readily deform than of a solid body that does not easily flow. The dark-band structure is composed of reddish and brown shades with irregular patches of gray, blue, and white clouds. The light zones are primarily yellow in color. The entire band structure is constantly undergoing changes in color and intensity. Clearly what we are viewing are clouds in Jupiter's atmosphere and not a solid surface such as the Terrestrial planets have. Most striking of all the atmospheric features is the Great Red Spot, which has been observed for at least 300 years. It is immense, being about four times the size of Earth.

In the early days of radio astronomy, Jupiter was found to be an intense source of radio radiation. If this radiation were just part of the planet's thermal radiation, then Jupiter would have to be extraordinarily hot. Since it is not hot, the radiation must be due to non-thermal processes, such as free electrons spiraling about magnetic lines of force.

Saturn's prominent rings make the sixth planet from the Sun, one of the most remarkable objects in the heavens. Brighter than all the stars except Sirius and Canopus, it shines with a steady ashen color. Saturn is second among the planets in mass and size. The mean diameter of Saturn minus its ring system is almost 10 times that of the Earth, and its mass is about 100 times greater. Its density is the lowest of any planet, 0.7 times that of water. The small mean density leads to the often quoted observation that if you could find a bathtub large enough, Saturn would float in it, being lighter than water. Rapid rotation, that is a rotation period of a little over 10 hours, and an unusually low density give it more polar flattening than any other planet, about 11 percent.

Saturn is twice as far from us as Jupiter, but the markings that we can see on the noticeably flattened disk of Saturn faintly resemble the banded cloud structure of Jupiter's atmosphere. The coloration is more restrained, and the details are less distinct. On rare occasions a bright spot may appear. Thus, as in the case of Jupiter, Saturn is a fluid-like body rather than solid, like a Terrestrial planet. As is true for Jupiter, astronomers have also detected weak radio emissions in low-frequency bursts from Saturn that are synchronized with its 10.2 hour rotation period.

Saturn's axis of rotation is inclined by $29°$ to its orbital plane. Since the plane of its rings is perpendicular to its rotation axis, the rings do not lie in the orbital plane and therefore present a varying aspect to Earth as the planet goes through its roughly 30-year orbital period. When seen almost edgewise, every 15 or so years, the rings almost disappear from sight, indicating that they are very thin compared to their radius. Most of Saturn's satellites orbit in the same plane as the rings, the planet's equatorial plane, and orbit outside the rings. Titan is one of the most massive satellites in the Solar System, and it is one that has been known for some time to possess an atmosphere.

On the night of March 13, 1781 William Herschel wrote in his observing journal:

"In examining the small stars in the neighborhood of H Geminorum I perceived one that appeared larger than the rest; being struck with its uncommon appearance...I suspected it to be a comet."

Herschel and other astronomers first believed the newly found object to be a comet and vainly tried to derive a cometary orbit for it. It was almost a year before they realized that this was a new planet.

Uranus has a radius four times larger than that of Earth and a mass almost 15 times greater. Although Uranus is somewhat larger than the more distant Neptune, it is less massive by about 15 percent. Consequently, it has a smaller mean density than Neptune but one that is larger than that of Saturn. Its average density is slightly higher than that of water. In a large telescope the slightly flattened disk is a light apple green in color.

Nearly 3 billion kilometers from Earth, Uranus presents an almost featureless appearance. Although a few atmospheric features have been reported, no extensive data on them exists. As with Jupiter and Saturn, we are probably seeing clouds in Uranus's atmosphere rather than a solid surface.

Uranus's rotation is peculiar in that its axis is tilted 98° to the perpendicular to its orbital plane, that is, it lies on its "side," so that we see it rotate in the reverse direction barely. For Uranus the retrograde rotation is due to the peculiar inclination of its axis, whereas for Venus it is a true reverse rotation. When its axis is along our line of sight every 42 years, half the sidereal period, we observe either its sunlit northern or southern hemisphere, while the opposite hemisphere is dark. One-quarter or three-quarters of its period later, 21 years or 62 years, its axis is at right angles to our line of sight, and we observe both the northern and southern hemispheres.

The reason for the axis of rotation to be lying nearly in the orbital plane of the planet is probably the result of several forces conspiring to push the planet into this state. Since the planet has most likely been in this state since shortly after its formation, we can only speculate that these forces would have been gravitational interactions among its components, collisions with interplanetary debris, and the retarding drag forces exerted by gas left over from the planet's formation.

The two brightest satellites of Uranus, Titania and Oberon, were discovered by William Herschel in 1787, only 6 years after he discovered the planet itself. In all, the planet has 27 known satellites. All the satellites move in nearly circular orbits that lie close to the equatorial plane of Uranus, the same plane as the ring system, but well outside the rings. In these respects, Uranus is similar to Saturn. The ring system was accidentally discovered in 1977 as a result of observations from an airborne telescope that was being used to re-measure Uranus's diameter and study its atmosphere as the planet passed over a background star. The expectations of future findings produced by such accidental events as the discovery of Uranus's rings contribute to the excitement and allure of science among not only the general public, but scientists as well.

After Uranus had been discovered accidentally, astronomers were long perplexed that even allowing for the perturbations of Jupiter and Saturn, Uranus's orbital motion was less predictable than that of other planets. The discrepancy was finally resolved in 1845 and 1846 by two astronomers, John Adams (1819-1892) in England and Urbain Leverrier (18ll-1877) in France. By a brilliant application of Newton's law of gravitation they arrived independently at the conclusion that there must be a disturbing body beyond the orbit of Uranus.

Leverrier's results were communicated to Johann Galle (1812-1919) of the Berlin Observatory, who received the information on September 23, 1846. Within half an hour after beginning his search, Galle located the new planet among a group of eight stars whose positions had been charted on a recently prepared map. Recent historical research suggests that Galileo probably saw Neptune in December of 1612 and January of 1613, fully 233 years before Galle found it, but he did not recognize that it was a planet. We know that Neptune passed extremely close to Jupiter, which Galileo was observing during that time.

Looking at Neptune through a telescope, we see a slightly flattened, bluish-green, almost featureless disk. Observers at times have reported irregular, indistinct markings and a bright equatorial zone, although observations of the planet are very difficult to make and subject to some degree of doubt.

Neptune's diameter is about 3.5 times that of Earth. Its mass is 17 times greater, and its mean density is one-third that of Earth. Neptune is about 30 AU from the Sun or 30 times Earth's distance. Thus the angular diameter of the Sun is one-thirtieth of what it is from Earth. For us the Sun has an angular diameter of $0.5°$, or 30 minutes of arc, so that from Neptune its angular diameter is 1 minute of arc. At a distance of 30 AU, Neptune's orbital, or sidereal, period is almost 165 years. Thus it has only completed a little over one orbit of the Sun since its discovery in 1846.

The largest of Neptune's 13 named

satellites, Triton, was discovered less than a month after the planet itself. It orbits Neptune in about 6 days in a direction opposite to the planet's eastward rotation. The orbital plane in which Triton moves is inclined to the equatorial plane of Neptune. Triton is a bit smaller than our own Moon, 2750 kilometers versus 3470 kilometers for the Moon. However, Triton's mass is only about 80 percent that of the Moon, possessing a lower mean density.

The smaller satellite, Nereid, takes nearly a year to swing around Neptune in a highly elongated ellipse, varying between about 1.5 million and almost 10 million km from the planet. Nereids's orbital plane is also inclined to Neptune's equatorial plane. Neptune's satellites are distinctly different from those of Uranus. Their inclined orbital planes and Nereids's elongated orbit continue to prompt speculations on their origin.

8.2. JUPITER AND SATURN

There are many aspects of a planet's atmosphere that astronomers want to know about, such as chemical composition, temperature, density, cloud composition, winds, and how these change with height, position over the surface, and time. Many of these details are not available for Earth's atmosphere, much less for the atmospheres of other planets. But from Voyager 1 and 2 we have learned a great deal about the atmosphere of the Jovian planets. Probably the most fundamental piece of information necessary for understanding a planet's atmosphere is the atmosphere's vertical temperature structure. On the way up through Jupiter's and Saturn's tropospheres, the measured temperature profile first declines and then increases into the stratosphere, where photons from the Sun can be directly absorbed.

The first constituents of Jupiter's atmosphere to be identified were methane and ammonia in the 1930s. Some 30 years later, the most abundant element, hydrogen, was identified and estimated to be 1000 times more prevalent than methane and ammonia. From these identifications, estimates for the hydrogen, carbon, nitrogen, and oxygen abundances indicate that

Jupiter's chemical composition, and similarly for Saturn, is more like that of the Sun than like that of the Terrestrial planets. In the 1970s and 1980s, primarily through infrared observations, several additional molecules were found to be minute constituents of Jupiter's atmosphere. Many of these molecules are probably also present in Saturn's atmosphere, but Saturn is colder than Jupiter, so that some compounds are probably frozen into solid crystals; thus they are not in a gaseous state capable of being observed spectroscopically.

Helium, the second most abundant element in the composition of the Sun, and presumably of Jupiter and Saturn, is not directly observable by spectroscopic means. Data from Pioneer and Voyager missions provide means for indirect determinations of the number of helium atoms per unit volume; the values derived, 10 percent for Jupiter and 6 percent for Saturn, are consistent with the solar composition hypothesis.

The most conspicuous aspect of Jupiter's and Saturn's atmospheres in visible light is their clouds. Knowing something about the atmosphere's vertical temperature profile and chemical composition provides clues as to what are the basic constituents of the clouds. For Jupiter and Saturn there appear to be three distinct cloud layers. The lowest layer is composed of water ice crystals or possibly liquid drops, the next of ammonium hydrosulfide crystals, and the highest of ammonia crystals. The middle one can also be thought of as a compound of the more elementary molecules ammonia and hydrogen sulfide. All the molecules forming the basic cloud particles should produce white particles, so other molecules must be responsible for coloring the clouds, which are red, yellow, brown, blue, and white. The most likely coloring agent is sulfur, which forms a variety of colored particles depending on molecular structure. This supposition has not been confirmed.

Infrared images of Jupiter and Saturn show that cloud color also correlates with altitude. Seen from outside, blue clouds lie at the deepest levels in the atmosphere and are visible only through holes in the upper clouds. Brown clouds are the next highest, above which lie white clouds, and finally, red clouds are the top layer. Compared to Jupiter, the greater spread in altitude for clouds in Saturn's atmosphere results from the smaller mass of Saturn, whose gravity is not as effective in compressing the atmosphere as is the more massive Jupiter.

The alternate light- and dark-colored cloud bands paralleling Jupiter's and Saturn's equator are constantly undergoing changes in color and intensity. Apparently this is because of the formation or dissolution of clouds of differing chemical compositions at different altitudes. There are the large-scale patterns, such as the bands themselves and the Great Red Spot on Jupiter, that last for years and sometimes centuries. This complex behavior betrays an involved atmospheric dynamics for both planets.

The dominant observable motions in the atmospheres are alternating eastward to the direction of rotation, and westward winds that correlate with the colored bands. Jupiter has five or six eastward and westward moving wind streams in each hemisphere, while Saturn has fewer but stronger ones. These winds are measured relative to each planet's rotation. In the case of Earth, there is only one low-latitude westward wind stream, known as the trade winds, and one mid-latitude eastward-moving jet stream. Jupiter and Saturn also have some vertical streaming.

Evidence suggests that these east-west winds have been constant in latitude and velocity for the last century or so. However, cloud bands with which they correlate are changing, as for example when small eddies between wind streams are sheared apart in 1 or 2 days. Eddies are deviations in what are otherwise alternating streams flowing east or west in the atmosphere. Where steady winds have velocities up to 100 m/s, eddy velocities are a few tens of meters per second.

Cloud motions on a small scale are by no means orderly. Voyager scientists were unprepared for the diversity and sometimes large turbulent motions in clouds observable in spacecraft photographs. Surprisingly, photographs failed to reveal cloud features smaller than about 100 km across. Narrow bands appear to coalesce and widen, while wide bands break apart. Material even seems to be transferring between bands.

Conspicuous in Jupiter's southern hemisphere is the oval-shaped Great Red Spot, measuring some 14,000 x 40,000 km. Although it has always been present since its telescopic discovery three centuries ago, it does vary both in size and intensity. Its sense of circulation, and that in other ovals in the southern hemisphere, is counterclockwise, whereas those ovals in the northern hemisphere rotate clockwise. This suggests that such ovals are high-pressure cells analogous to those in the Earth's atmosphere. Small white clouds can be seen circulating around the Great Red Spot over periods of a week or so, whereas by comparison, the interior is relatively calm.

Saturn also has oval-shaped circulation cells in its atmosphere. A brown oval was first photographed during the August 1981 flyby of Saturn by Voyager 2. It is not known whether the eddies and ovals on both Jupiter and Saturn extend as deep into the planet as do the wind streams. However, the long-term persistence of the winds and the short life for eddies and ovals are probably related to the mass of material involved in the phenomena. Thus the winds probably extend deep into the planet, while the shorter-lived eddies are relatively shallow structures. However, this is still quite speculative.

Winds on Earth draw their energy from unequal heating by the Sun between the equator and the poles, and in general, the temperature decreases poleward by some 30K at almost all levels in the atmosphere. Even though the Sun heats the equatorial regions of Jupiter and Saturn more than it does the polar regions, just as it does on Earth, on Jupiter the temperature difference between the equator and the poles is no more than 3 K. Thus some mechanism must transport heat from the interior of the planet into the polar regions, reducing the temperature difference.

8.3. URANUS AND NEPTUNE

In the 1930s, spectroscopic studies revealed methane in the atmospheres of Uranus and Neptune, as in those of Jupiter and Saturn. Since then, hydrogen has been identified, helium has been inferred indirectly, and some other hydrogen-containing molecules have also been discovered. Greatly distant from the Sun, Uranus and Neptune are very cold, and thus a number of molecular combinations are probably frozen into a crystal or liquid-drop form.

Important for Uranus is the fact that its axis of rotation lies almost in its orbital plane, causing regions near the poles to remain alternately in sunlight or darkness for periods approaching 42 years. What affect such a phenomenon has on the overall structure of the atmosphere and how much of a difference it produces between Uranus and Neptune is only now being considered.

Although alike in many respects, ground-based and Voyager 2 evidence suggests that the atmospheres of Uranus and Neptune are not highly similar to each other or to those of Jupiter and Saturn. On January 24, 1986, Voyager 2 made its closest approach to Uranus, coming within 80,000 km of the planet. The striking aspect of the planet as seen from the advantageous position of Voyager 2 or from the distant Earth is how bland and featureless the blue-green planet appears. Prior to Voyager 2, ground-based data had been interpreted to say that there were no clouds in the Uranian atmosphere unlike Jupiter's and Saturn's atmospheres. Although Voyager 2 did finally confirm the existence of clouds in the planet's atmosphere, they are considerable smaller than the planet's diameter and only about five percent brighter than the background atmosphere. Icy materials, formed from hydrogen, carbon, nitrogen, and oxygen, appear to be the principle constituents of Uranus, and at the very low temperatures in the Uranian atmosphere these compounds condense to form clouds of ice crystals. Methane freezes at the lowest temperatures, so that the top cloud layers are probably composed of methane ice crystals. These methane clouds are probably extensive enough to obscure the underlying ammonia and water clouds. This would explain why in infrared spectra of the planet there are no signatures of these two molecules. Finally, the blue-green cast to Uranus is due to selective absorption of the reddish sunlight by methane molecules in its atmosphere.

On Earth the heating of equatorial zones

and the consequent temperature decline toward the poles produces the strong eastward moving jet streams at mid-latitudes. But on Uranus sunlight comes at times almost down the rotation axis into the polar regions. Thus it was questionable whether or not Voyager 2 would find wind patterns similar to those in the atmospheres of Jupiter and Saturn. Voyager evidence, however, shows that indeed there is east-west type of winds in the Uranian atmosphere. The amount of solar radiant energy arriving at Uranus is so weak in comparison to that at the Earth's distance that the winds may not be caused by unequal heating as in the case of Earth.

In contrast to Uranus, the atmosphere of Neptune appears to possess a variable haze or extensive clouds of unknown chemical composition as seen from ground-based telescopes. At times nearly half the planet's atmosphere is hazed over. This haze can dissipate and re-form in a matter of weeks or even a few days. The haze is partly responsible for trapping solar radiation, so that Neptune's upper atmosphere is warmer than that of Uranus. Of course when

Voyager 2 arrives at Neptune in August of 1989, we will undoubtedly have to modify our understanding of Neptune's atmosphere as we are still in the process of doing for Uranus as data is evaluated and analyzed.

8.4. INTERNAL STRUCTURES OF JOVIAN PLANETS

The general comments made earlier about models of planetary interiors are applicable to the Jovian planets as well as to the Terrestrials. Larger masses, and the fact that the Jovian planets contain far more easily vaporized materials than do the Terrestrial planets mean that the internal structures of the Jovian planets are not like those of the Terrestrials. Jupiter and Saturn are the only planets composed primarily of hydrogen and helium, as is the Sun. Only hydrogen and helium could give Jupiter and Saturn their mean densities of 1.31 and 0.69 g/cm^3, respectively, for the temperatures and pressures that characterize each planet. However, the masses of Uranus and Neptune are 5 and 6 percent, respectively, that of Jupiter, while their mean densities are about equal to or larger than Jupiter's. This indicates that whereas Jupiter and Saturn are composed primarily of hydrogen and helium, the percentages of carbon, nitrogen, oxygen, and possibly silicon, and iron in Uranus and Neptune must be greater than in Jupiter. That is, Uranus and Neptune do not have solar compositions, but rather they have larger fractions of icy and rocky materials in their composition.

The rapid rotation of Jupiter and Saturn, coupled with their composition of low-density materials, argues that their internal structures are more fluid than solid. Another significant factor is that Jupiter and Saturn give off more heat than they receive from the Sun. In the case of Jupiter, the heat given off is about 1.5 to 2 times the amount received from the Sun, and for Saturn, it is between 2 and 3 times the amount. Hence Jupiter and Saturn have internal sources of heat. It is extremely unlikely that the heat source is anything as exotic as that in the Sun and the stars; Jupiter and Saturn are not small stars. But it is fair to say that they are more like the Sun than like the Earth, and they are clearly an intermediate type of body. The internal heat source probably results from the conversion of gravitational potential energy into thermal energy as the two planets contracted during their formation and after. In fact, it is likely that they are still contracting, but very slowly.

Both Jupiter and Saturn have dense cores of rocky and icy materials, rather than compressed hydrogen and helium. The core is about 4 percent of the mass of Jupiter and 25 percent of the mass of Saturn, with temperatures in the range of 20,000 to 30,000 K and densities ranging from 10 to 20 g/cm^3. Surrounding the core is a layer existing under a pressure in excess of 3 millions times Earth's atmospheric pressure. In it hydrogen and helium behave more like liquid metals than solids. The upper boundary of the metallic liquid zone is rather abrupt, giving way to a molecular liquid mantle of hydrogen and helium. Through both the metallic and molecular liquid zones, which are 96 and 75 percent, respectively, of the masses of Jupiter and Saturn, the temperature and density decrease. The molecular liquid mantles gradually change to molecular gases, which are then the atmospheres of the two planets.

Like Jupiter and Saturn, Uranus and Neptune have a three-layered structure, but unlike the Solar System giants, each layer is of quite different chemical composition. The core of each planet is probably a rocky, mostly iron and silicates, and icy material, principally methane, ammonia, and water. For Uranus, the pressure of overlying layers may not be sufficient to make the core solid, but it remains a thick, viscous liquid with convective motions in it. However, Neptune's greater mean density suggests that its core is solid.

Surrounding the core of each planet is a liquid mantle of water, methane, and ammonia, in which there may be some convective motions for Neptune but not for Uranus. Finally, each planet has a thick crust of hydrogen and helium that is compressed by gravity into a very dense gas. The crusts gradually give way to low-density atmospheres. Thus, like Jupiter and Saturn, these planets have no solid surface surrounded by a thin atmosphere as the Terrestrial planets have.

Calculated models for the interiors of

both planets suggest that their central temperatures are on the order of 7000 K. Since Jupiter and Saturn emit more radiant energy than they receive from the Sun, is it not likely that the same is true for Uranus and Neptune? Yes, one might well expect this to be the situation for both. But Voyager 2 data suggests that Uranus has lost most but not all of its internal heat since it was formed some 4.6 billion years ago. As much as 30 percent of the heat radiated by the planet may come from its deep interior rather than from the Sun. For comparison the comparable figure for Earth is about 0.01 percent. The strength of Uranus' internal heat source is an important clue to its past evolution. For Neptune, it appears to be radiating about twice as much heat as it receives from the Sun. Why this difference between Uranus and Neptune which should be reasonably similar bodies, is really not completely known as yet.

8.5. JOVIAN MAGNETOSPHERES

Jupiter is the strongest radio emitter in the Solar System after the Sun, emitting both thermal and non-thermal radiation. At times its radio emission exceeds even the Sun's in intensity. The non-thermal radiation is a type of synchrotron radiation, and it results from Jupiter's having a magnetic field and energetic free electrons in radiation belts that spiral around the magnetic field lines. These radiation belts are analogous to Earth's Van Allen belts. There are occasional bursts having energies up to 10^{17} erg/s. The bursts are more intense when the nearest Galilean satellite, Io, appears on one side of Jupiter as viewed from Earth. Why should the position of Io make a difference? We suspect that it is due to the motion of Io through Jupiter's magnetic field, disturbing the field and the electrons trapped in it.

Pioneer space probes ran into the bow-shock wave formed by the solar wind's interaction with Jupiter's magnetic field as far out as 108 Jupiter radii. Data from the two Pioneer craft and the two Voyagers indicate that the boundary of the magnetosphere in the direction of the Sun varies between about 50 and 100 Jupiter radii. The planet's inner radiation belt is like Earth's Van Allen belts but from 5000 to

10,000 times more intense.

Farther out, the magnetic field flattens into a disk that extends several million kilometers from the planet, and its long tail, flowing out opposite to the direction of the Sun, extends an unknown distance beyond the orbit of Saturn. The shape is influenced by the large centrifugal force that results from the planet's rapid rotation.

Saturn's magnetic field also defines a zone about it, or a magnetosphere, in which it can control the motions of subatomic particles. The Saturnian magnetosphere is intermediate in size, and its intensity lies somewhere between that of Jupiter and Earth. All three are based on a common framework of physical principles, but each possesses its own distinctive character.

Prior to the late summer of 1979, astronomers could only speculate on the magnetic field and radiation belts around Saturn. During that summer, Pioneer 11 detected the boundary of the magnetosphere laying some 24 Saturnian radii from the planet; its rings extend about 6 radii from the planet. Saturn's magnetosphere is apparently more disk-like than that of the

Earth, which is more spherical but less so than Jupiter's larger magnetosphere.

Finally, we find that Uranus and Neptune possess magnetic fields that produce magnetospheres around themselves as do Jupiter and Saturn. Voyager 2 detected a magnetic field extending around Uranus. The magnetic axis is not aligned with the planet's axis of rotation but is inclined by an angle of about 60°, so that the magnetic field wobbles significantly as the planet rotates. Also Uranus' magnetic field is not as simple a field as had been expected before Voyager 2's arrival. Thus there are many puzzles about the magnetic field and magnetosphere for Uranus yet to be explained.

8.6. RING SYSTEMS OF JOVIAN PLANETS

Of all the aspects of the Jovian planets, their ring systems are among the most captivating. Galileo first observed what we know as Saturn's rings in July of 1610, but it was not until 1655 that Christian Huygens proposed that they are a flattened disk of matter detached from the planet. In 1857,

James Clerk Maxwell showed mathematically that they must consist of numerous tiny bodies in orbit around Saturn. This was experimentally demonstrated in 1895 from Doppler shifts that showed that each of the ring particles pursues its independent orbit around Saturn in accordance with Kepler's third law. The farther out from the planet, the lower are the particles' speeds, whereas a solid ring would rotate fastest at the farthest point from the planet.

It can be shown with reasonable mathematical precision that particles swarming around a planet eventually form a thin system of rings in the equatorial plane. A ring system is produced by the gravitational attraction of the planet and many gravitational interactions of ring particles with each other. Satellites of the planet play an important role in sculpting the appearance of the rings and in keeping them from spreading out in the equatorial plane. In addition, the ring system forms within several planetary radii of the planet's surface and is not able to form at greater distances. Although the same basic principles underlie the three known ring systems, Saturn's rings are much more elaborate and complex than those of Uranus and Jupiter.

Three concentric rings have been known for some time and are labeled A, B, and C in order of decreasing distance from Saturn. The brightest ring, B, is separated from the somewhat fainter ring A by a dark space of about 5000 km, called Cassini's division. The faintest of the three major rings, the so-called crepe ring C, lies inside the inner edge of ring B. An exceptionally faint ring D, which lies inside the inner edge of ring C, has been found by Voyager investigations. Outside ring A, other faint rings, known as E, F, and G, have been identified. The vertical extent of all the rings is less than a couple of kilometers, possibly even as thin as 100 m. Given their immense diameters, they are proportionally speaking thousands of times thinner than a razor blade compared to its dimensionality.

The three major rings, A, B, and C, lie within the critical distance called the Roche limit, which is equal to about 2.4 Saturnian radii. This limit is named after the nineteenth-century French mathematician

Edouard Roche (1820-1883), who found that inside this limit the gravitational attraction exerted by a planet on two adjacent orbiting particles is larger than the attraction of the two particles for each other. Whether the rings were formed inside the Roche limit by the breakup of a satellite, comet, or other body or whether Saturn's gravitational force prevented primordial particles from coalescing to form a satellite is unknown.

High-resolution photographs made by the Voyager spacecraft, surprised astronomers when they revealed that the three major rings, A, B, and C, are made up of hundreds, if not thousands, of very narrow ringlets. It had been thought that Cassini's division appears dark because it is devoid of any particles. However, photographs taken from the backside of the rings show that even Cassini's division is crammed with something like 100 ringlets. The Voyagers even provided evidence that some ringlets are not circular. In addition, ring F, has illusory knots, braids, and twists in it, which had not been previously predicted from the computer models.

The nature of the ring particles is

suggested by the way sunlight bounces off of them, a process called scattering. The relative amounts of light scattered in various directions by the ring particles depends on both their sizes and the wavelength of light. From data collected by the Voyager spacecrafts, scientists estimate that the ring particles vary in size from a few microns up to a few tens of meters. But the most abundant particles are snowball-sized particles around 10 cm. The range of particle sizes in each of the rings, however, does not appear to be the same. For example, ring C and Cassini's division appear to contain relative few of the smaller particles. Apparently, the larger-sized particles in Cassini's division do not scatter photons in the backward direction as well as smaller particles do, so they appear dark from the sunlit side but bright from the backside.

From studies of infrared data obtained by the Voyager spacecrafts, it appears that the ring particles are water ice or at least covered with water ice. In addition, there are subtle differences in color between rings and even ringlets, which suggest that some other substance or substances are mixed with the water ice. These trace substances have not yet been identified. The existence of the color variations between rings and its persistence suggests that ringlet particles do not migrate between rings.

Probably the most unexpected aspect found by the Voyagers was wedge-shaped spokes orientated radially out from the planet in ring B. From the sunlit side, the rings appear dark. However, looking back toward the Sun, they appear bright. They are perplexing in that if they are produced somehow by the ring particles, Keplerian motion should dissolve the spokes in a relatively short time. However, they are seen to last close to 10 hours. Studies of light scattered from the spokes suggest that they are fine dust-like particles that are situated a few tens of meters above the rings. Nevertheless, there are still some puzzling aspects to the spokes for which we have no completely satisfactory solution.

The notion that Jupiter possesses a ring system like that of Saturn was proposed some 40 years ago. Pioneer 11 data were interpreted as consistent with the existence of a system of tiny satellites forming a ring about Jupiter. This was at best speculation, and it was only Voyager 1's photograph of the Beehive star cluster that finally revealed the ring system.

Data from the Voyagers reveal a ring system composed of three parts. The primary ring, and brightest part of the system, starts abruptly at 1.81 Jupiter radii and ends, while fading gradually, 6,000 km closer to the planet. At most this primary ring is about 30 kilometers thick. There is also a tenuous sheet of material that is several times fainter than the bright ring extending from the primary ring smoothly toward to the planet's outer atmosphere. Surrounding the ring and sheet is a faint, lens-shaped halo some 20,000 km thick close to the planet. Saturn's rings are not embedded in such a halo.

The particles composing the ring appear to be smaller on the average, typically a few microns in size, than Saturn's ring particles. Also unlike Saturn's ring particles, those of Jupiter's and Uranus's ring systems are quite dark. Thus they are not water ice or coated with water ice. Evidence suggests that they are probably silicate particles whose origin is unknown.

Occasionally, a planet will pass between the Earth and a star. Such an event is called an occultation, from the Latin word meaning "hiding." In recent years, astronomers have carefully monitored these occultations because the time and place on Earth at which the occultation will be visible can be calculated. It requires a precise knowledge of the planet's orbit to make such a calculation, and the precision with which the prediction is confirmed by the observation in turn tells us how well we really know the planet's orbit.

As the planet begins to occult the star, its atmosphere, which is partially transparent, covers the star first, so that there is a gradual dimming of the star. If there were no atmosphere, the star's brightness would remain constant until the opaque body of the planet cut off all light; the change would be sudden, not gradual. In this manner, astronomers aboard the Kuiper Airborne Observatory, an airplane fitted with an infrared telescope, flying high over the Indian Ocean discovered a ring system around Uranus on March 10, 1977. About a half hour before the occultation was to take place, the star's light dimmed unexpectedly for a few seconds, followed by four other dips in brightness minutes later. The sequence was repeated in reverse as the star passed beyond the disk of Uranus on the other side. Since the original discovery of five rings, four less prominent rings have been identified, making a total of nine rings.

Voyager 2 confirmed that the Uranian rings are quite narrow, not more than a few up to one hundred km in width and are separated from each other by hundreds of km of virtually empty space. The radii for all the rings lie between 1.6 and 1.95 planetary radii. The rings are not all circular, but they are dark with sharp edges. Not all the rings lie in the planet's equatorial plane, although they are close. Thus the rings are orientated almost perpendicular to the planet's orbital plane. Hence the origin of the rings is closely related to that of Uranus, since the planet's equatorial plane is almost perpendicular to its orbital plane. In addition to observing the nine known rings, Voyager 2 also discovered one new ring that is quite narrow and faint and about 100 almost transparent bands that are virtually invisible from the Earth.

The particles composing the known rings are larger than had been anticipated, being from about 10 cm size up to several meters. In contrast the particles composing the newly discovered dust bands are quite small, being nearly microscopic in size, roughly 0.02 mm. In addition to the new ring and dust bands, ten small satellites were found all of which lie closer to Uranus than the closest of the previously known satellites Miranda. Two of the new satellites are quite close to the rings. But the most interesting discovery is that the main rings and the new satellites are charcoal black. Photographing them against a black background was a monumental accomplishment. Since the ring particles are poor reflectors, they can not be coated with water (or ammonia or methane) ice. More likely, they are a silicate- or carbon-bearing material.

Using occultations of background stars by Neptune, evidence accumulated suggesting that this planet possesses a ring, or part of one. The best data prior to Voyager 2's flyby implied that the supposed ring is just 10 to 20 km wide, situated in Neptune's equatorial plane, and is roughly 76,000 km, or about 3.1 Neptune radii, from the center of the planet. To explain a lack of symmetric behavior in the occultation measurements on opposite sides of Neptune, it had been suggested that the supposed ring is narrower on one side than the other, or it is fragmented and discontinuous, with most of the ring material concentrated in bands along the ring's perimeter. The definitive data on the ring system's existence, however, came in August of 1989, when Voyager 2 made its flyby of the planet.

Chapter 9
Minor Bodies and the
Interplanetary Medium

We have made several references to the fact that within each category of Solar System bodies, there appears to be a large variation in the sizes of these bodies, including the equivalence in size of bodies in different categories. For example, there are bodies called planets, such as Mercury, that are equivalent in size to some of the satellites of Jupiter, such as Ganymede and Callisto. Another example is that there is no comparison in size between Jupiter and Mercury and yet all both are called planets. Jupiter is vastly larger than Mercury. What this means for us is that our natural inclination to believe that bodies belonging to the same categories must be alike in either size, composition, appearance, or in some sense, is not applicable to Solar System bodies. For example, we can not think about the various categories of Solar System bodies as a hierarchy of different sized objects, such that planets are always larger than satellites, which in turn are always larger than asteroids, and so forth. Not only is this true about size, but it is also true for chemical composition and appearance. The Terrestrial and Jovian planets are quite different in these two aspects. The categories of Solar System bodies that we have heard about since childhood are historical accidents based on what mankind thought they were discovering. Thus we must be careful to pay special attention to the diversity that exists in each category, which is especially true for the minor bodies of the Solar System, that is dwarf planets, satellites or moons, asteroids, meteoroids, and comets. We start first with Pluto, once called a planet, now a dwarf planet.

9.1. PLUTO

Although Pluto is not one of the Jovian planets, it does orbit in the outer part of the Solar System. However, it is neither a Jovian planet nor for that matter a Terrestrial planet. In fact, until 2006, when a meeting of the International Astronomical Union was adjourning, Pluto was called a planet. It is now regarded as a "dwarf planet."

Spurred by the success of the discovery of Neptune in 1846 as noted in the previous chapter, astronomers searched for evidence for even more distant planets. Percival Lowell was convinced by his calculations begun in 1905 that minute discrepancies still complicated the orbit of Uranus. Neptune had not been observed long enough to provide useful data. He concluded that the irregularities might be caused by a planet beyond Neptune.

Several years of intermittent and unproductive search passed. Then in January of 1929, the Lowell Observatory acquired a 13-inch photographic refractor and put a young assistant, Clyde Tombaugh, to work on a new search. After a year of photographing star fields along the ecliptic and later all over the sky, Tombaugh made the historic find on photographs taken in January of 1930.

Pluto's great distance from the Sun, with a semi-major axis of almost 40 AU, and a long sidereal period of almost 248 years, have made it a difficult planet to study. Pluto's large eccentricity carries it as close to the Sun as about 30 AU and as far away as almost 50 AU, a variation of almost 20 AU, or about 3 billion kilometers. Thus during a portion of its orbit it is closer to the Sun than Neptune is. In fact, Pluto was closer to the Sun than Neptune from the winter of 1978 until the spring of 1999. Pluto is now moving along that portion of its orbit that is north of the plane of the ecliptic. Pluto reaches aphelion while it is south of the ecliptic plane. At that point it will be almost 14 AU below the plane of the ecliptic, which is a greater distance than Saturn is from the Sun. Pluto's crossing of Neptune's orbital plane is done well above or below it so that there is no chance for the two planets to collide.

At the time of its discovery it was estimated that Pluto was approximately Earthlike in size and mean density. Thus there was the possibility that a Terrestrial planet had been found in the outer Solar System. Since then, with a longer period for study, estimates of Pluto's mass and radius have decreased substantially. It is by means of Pluto's gravitational attraction for Neptune and, to a lesser extent, Uranus that astronomers tried to estimate its mass. However, it was not until the discovery of Pluto's moon Charon, that a value for its mass could better be determined. Our best estimates at this time suggest that Pluto's mass is too small to have ever produced the effects on Uranus' orbit that presumably lead to its discovery. Thus in some sense the discovery of Pluto is owed to the systematic search made by Tombaugh and not to the predicative powers of Newtonian gravitational theory.

The planet's small disk, measured with difficulty even by a large ground based telescope, is less than 2/10 Earth's diameter. Pluto's mass is much less than one percent that of Earth's. For comparison Pluto's mass is just 1/500 of Earth's mass, while Mar's mass is 1/9 of Earth's, Mercury's 1/18 of Earth's, and the Moon's 1/80 of Earth's. Pluto is an icy body because of its low mean density at around 2 g/cm^3. This supports the belief by astronomers that Pluto is composed primarily of water in solid form, or a mixture of water and other ices, such as methane and ammonia. A suggested structure for Pluto is a small rocky core covered by an extensive water ice mantle with a crust of methane ice.

Pluto's brightness varies slightly, presumably because sunlight is reflected unevenly from its surface, perhaps because of lighter and darker areas. Photoelectric observations of these variations reveal that the period of Pluto's rotation is 6.4 days. The nature of the brightness variation also suggests that Pluto's axis of rotation is highly inclined relative to its orbital plane. Pluto's satellites orbit in the equatorial plane of the planet, thus Pluto's axis of rotation is almost in its orbital plane, like that of Uranus, or about 120° off perpendicular.

Infrared observations suggested a surface composition dominated by ices, primarily methane. Pluto's low surface gravity and extremely low temperature mean that its atmosphere is a tenuous which will probably completely freeze out as Pluto gets further away from the Sun, and approaches aphelion. However, its observed reflectivity is not consistent with a prediction of methane ice, or frost, alone. That is, some darker patches may possibly be a rocky silicate material.

Pluto has five known satellites, Charon, Hydra, Nix, Kerberos, and Styx. Charon was first photographed in June of 1978. The discovery was made during the examination of photographs taken as part of the routine task of refining data on the planet's orbital motion. That image of Pluto was elongated, with a bulge at the top that was the unresolved image of Charon. Although such photographs do not resolve the satellite, Charon's existence was again confirmed by observing a sequence of its eclipses of Pluto, which became visible on Earth in 1985. From these eclipse observations, ground-based speckle photographs, and images from the Hubble Space Telescope Charon's diameter was estimated around 1200 km.

If Pluto and Charon have equal mean densities, the satellite is about 10 percent the mass of Pluto, which would make it by far the largest satellite in the Solar System in comparison with its parent planet, the Moon being about 1.2 percent of the Earth's mass. Charon orbits Pluto in an approximately circular orbit inclined to that of Pluto, at an estimated distance of 19,400 km and with an orbital period of 6.4 days.

If you think astronomers were surprised by the discovery of Charon, they were even more surprised to discover that Pluto has at least four other moons. Pluto's moons Nix and Hydra were discovered in 2005. Kerberos was discovered by the Hubble Space Telescope in 2011, and Styx in 2012. Thus, Pluto, a dwarf planet, has at least five natural satellites or moons, and may even have a ring system of its own. We shall all learn more about this peculiar celestial body when the New Horizons spacecraft encounters Pluto in the summer of 2015, if all goes well.

Is it possible that even more remote bodies lie beyond Pluto? There is no physical reason that other objects similar to Pluto cannot exist. In fact, for 13 years after the discovery of Pluto, Tombaugh continued to search for more trans-Neptunian objects. While Tombaugh was not successful in his efforts, more modern attempts have proven more fruitful. Today there are some twenty named trans-Neptunian objects. The largest of these are Sedna, Haumea and Eris, discovered in 2003, and Makemake discovered in 2005.

9.2. SATELLITES OF TERRESTRIAL PLANETS

Only three satellites are known for the Terrestrial planets. One of course is our own Moon. The other two are the little satellites of Mars, which were discovered in 1877. Phobos, the inner one, and Deimos, the outer one, are potato-shaped bodies with cratered surfaces. Phobos orbits eastward, just as our Moon does, and in the same direction that Mars rotates, in a period of 12.5 hours at a distance of about 6000 km from the surface of Mars. This gives it an angular size, as seen from the surface of Mars, of about half that of our Moon, or about 15 arc minutes. Since it revolves around Mars much faster than the planet rotates, it rises on the western horizon and sets on the eastern horizon 5.5 hours later. As observed from its primary, Phobos is the only natural satellite in the Solar System whose motion is of this form.

Phobos is about 27 km long, 21 km high, and 19 km wide. Both Phobos and Deimos have been shaped by high-velocity impacts, which appear to have sheared off large sections of each satellite. In addition, both have many craters, but no significant amount of ejected material or craters with central peaks, a reasonable result since their gravitational attraction is very small. Some of the material ejected during cratering simply escapes and does not fall back to the satellites' surfaces.

Phobos seems more heavily cratered than Deimos, the largest crater, Stickney, being about 10 km across. It also has mysterious long parallel grooves across a large part of its surface. They are a few hundred meters wide and a few tens of meters deep and may have been formed by the same impact that caused the crater Stickney.

Deimos is about half the size of Phobos, some 15 km by 12 km. It orbits Mars some 20,000 km from the planet's surface in a period of 30.3 hours. Its angular size, as seen from Mars, is quite small, roughly equivalent to a quarter viewed at a distance of about 40 m. Its orbital period is somewhat longer than the rotational period of Mars, so it rises on the eastern horizon and sets on the western horizon nearly 3 days later, while going through its phases twice.

The darkness of the surface of both satellites is probably due to carbon- and water-rich minerals, such as are found in black, crumbly meteorites known as carbonaceous chondrites. A number of asteroids appear to have similar surfaces. This has led to the speculation that Mars's satellites may be captured asteroids acquired early in the planet's life.

9.3. ASTEROIDS AND METEOROIDS

On January 1, 1801, a Sicilian astronomer, Giuseppe Piazzi (1746-1826), accidentally discovered a faint object whose orbital motion was that of a body 2.8 AU from the Sun. Although Piazzi thought it was a comet, others noted that it was located about where a major planet would be expected according to the old Bode's Law. The object was later named Ceres, after the Roman goddess of agriculture. Shortly afterward, three more objects were discovered with orbits near 2.8 AU: Pallas in 1802, Juno in 1804, and Vesta in 1807.

Since photographic techniques were introduced into astronomical research in the 1890s, more than 3000 of these bodies have been discovered.

So instead of one planet in the slot at 2.8 AU, many small bodies orbit in the region between Mars and Jupiter. William Herschel called these objects asteroids because in a telescope they looked like stars. Better than 90 percent of them have orbits in an asteroid belt between 2.2 and 3.3 AU, with periods from 3 to 6 years. For several asteroids, their orbits are more elliptic than are those of the planets and more inclined to the ecliptic. However, they move in the same direction as the planets around the Sun.

Asteroids vary in size from Ceres (960 km) down to an estimated 1 million that are more than 1 km in diameter and countless numbers of even smaller ones. All the asteroids together add up to no more than six ten-thousandths of Earth's mass. Ceres alone constitutes about 20 percent of the mass of all the asteroids.

Photometric studies of asteroids have for some time been interpreted as showing that they differ in size, shape, and rotation.

All but the largest ones are too small to show a measurable disk. From the variations in their brightness, it has been assumed that most have somewhat irregular shapes with periods of rotation measured in hours.

Their colors put nearly all asteroids into two categories: Some are bright reddish or sandy-colored, a sign of iron and magnesium silicates, and they populate mostly the inner part of the asteroid belt. Most asteroids, however, have the darker neutral color of material called carbonaceous, that is, containing various carbon- or water-rich compounds and occupy mainly the outer part of the belt.

Collisions between two asteroids may produce effects ranging from craters, if a small one collides with a large one, to fragmenting of both asteroids, if they are of comparable size. For example, if the body producing crater Stickney on Mars' satellite Phobos had been a little larger, Phobos might have been broken into many small pieces. As it is, the grooves on Phobos may be large cracks produced by the impact.

In October of 1977, the asteroid Chiron was discovered traveling in a highly eccentric orbit, with an eccentricity of 0.38, at an angle of 6.9° to the plane of the ecliptic. It ranges between 8.5 and 18.9 AU, or roughly between the orbits of Saturn and Uranus, with a period of 50.7 years. Although a couple of other asteroids are known to have aphelion points near Saturn's orbit, Chiron is the most distant yet discovered. Because of its great distance from the asteroid belt, there is a question whether it might be the first discovery in an outer zone of asteroids. Or possibly it confirms the suspicion that asteroids and comets are objects related to each other, although not having the same composition. Together they are all that remains of the small bodies, called planetesimals, which once filled the Solar System and ultimately coalesced to form the planets 4.6 billion years ago.

There are at least a thousand asteroids with diameters greater than 1 km that at some time during their orbit come close to Earth. Eventually, some of these will collide with Earth, as other asteroids have done in the past, for example, the 35-m-diameter one that caused the Barringer Meteor Crater.

Current estimates are that, for those with diameters under 0.1 km, one will collide with Earth every 10,000 years, releasing as much energy as 1000 atomic bombs, or about 13 million tons of TNT. An asteroid with a diameter close to 10 km should strike Earth once every 100 million years with an energy equivalent to 1 billion atomic bombs, or about 13 trillion tons of TNT. In fact, it has been proposed that the well-documented extinction, the so-called Cretaceous-Tertiary Extinction, of approximately 60 percent of all animal species that occurred about 65 million years ago was caused by climatic changes from the dust kicked up by the impact of a 10-km asteroid. So massive was the dust that was ejected into the atmosphere that a layer of clay 1 to 2 cm thick was deposited worldwide of which about 7 percent was carbonaceous matter from the asteroid.

As much as 1000 tons of cosmic debris, billions of microscopic particles, pepper the Earth daily. We are aware only of those weighing a significant fraction of a gram, which produce the so called shooting stars that flash across the sky. All but a few are too small to leave luminous trails. These solid particles are called meteoroids before they encounter Earth. Those large enough to survive flight through our atmosphere and land are called meteorites. And the luminous trails of the smaller particles that are completely vaporized in the atmosphere are called meteors. The largest and brightest meteors are referred to as fireballs or bolides.

Our atmosphere slows incoming meteoroids and transforms their kinetic energy into radiant and thermal energy. A meteoroid passing through the atmosphere leaves a wide, dense column of electrons stripped from atoms and molecules in its path. As ionized atoms regain their electrons, they de-excite, emitting photons that make the momentary luminous trail we see from the ground as a meteor or shooting star. Anything that remains of the vaporized meteoroid slowly filters down through the air as dust and solidified droplets of melted meteoroid.

When meteoritic particles strike the Earth, they are moving anywhere from about 10 up to 72 km/s, depending on the angle at which they encounter Earth. The velocities convince us that meteoroids belong to the Solar System, moving in independent orbits around the Sun.

The normal observed rate for meteors is about 10 per hour over the entire sky. Fewer meteors are seen before midnight than after midnight. This is so because during the evening hours we are on the back side of Earth, facing the direction opposite to Earth's orbital motion, and we see only the swift meteoroids overtaking us from the rear. During the morning hours, Earth's rotation has turned us so that we are facing in the same direction as its orbital motion. Hence we see those meteoroids that we overtake and those which meet us head on.

Several times during each year we see meteor showers, swarms of shooting stars darting from a small area in the sky. These showers can persist for hours or days. On such occasions, Earth is passing through a large group of particles moving in ribbon-like fashion along an orbit around the Sun. Perspective makes their tracks seem to diverge from a small spot in the sky called a radiant. Such showers are named after the constellation in which the radiant appears.

About 100 years ago astronomers found that some meteoroids travel in orbits much like those of some comets. They had found a link between meteor showers and short-period comets (to be discussed later in this chapter). The particle swarms are the resulting debris left after the evaporation of a comet. For example, on the night of November 13, 1833, watchers in the southern part of the Atlantic seaboard were awestruck as over 100,000 shooting stars per hour plummeted from the constellation Leo for 3 hours. The great display was produced when Earth encountered a swarm of meteors orbiting the Sun in a period of 33 years and associated with comet Tempel (1866 I). The comet itself has long since vanished, leaving the meteor shower as a remainder of its existence. The subsequent meteoric displays of 1866, 1899, and 1932 were progressively weaker; then on November 17, 1966, a fairly spectacular meteor shower was observed in the southwestern part of the United States. Other peaks in the meteor shower display took place in November of 1999, 2001 and 2002.

With the passage of time, a meteor stream, which is made up of conglomerates of fine dust and ice-covered particles, is strung out along the comet's orbit. This ribbon of particles typically averages about 50,000 km in cross section. Thus Earth must come fairly close to a meteor stream, therefore, in order for us to see a meteor shower.

Most meteorites are discovered accidentally years after they fall. Of some three dozen recovered meteorite falls weighing more than a ton, only a few were seen descending. Not many falls are ever recovered. Most meteorites land in the oceans or in unoccupied places, where their fall is less likely to be observed. No known record tells of a community destroyed or an individual killed by a meteorite despite some close calls. Approximately 3000 meteorite specimens have been recovered and catalogued for study.

Meteorites striking Earth have probably formed thousands of craters, but only 200 or so have been found. One major impact in 10,000 years is a reasonable estimate, and at that rate at least 50,000 giant meteorites must have struck Earth in the past 500 million years. But the fossil craters left by many of these may lie buried and unnoticed in the Earth's crust. Probably most of them have been obliterated by weathering, erosion, and geologic processes.

One that we know about, near Winslow, Arizona, is the Barringer Meteor Crater, created by a meteorite weighing at least 30,000 tons. It struck the Earth about 25,000 years ago and must have devastated all plant and animal life within a large area. The crater is over 1.2 km across and 167 m deep. Thirty tons of shattered iron fragments have been picked up within about 6 km of the crater.

At 7 A.M. on June 30, 1908, a tremendous fireball flashed across the sky in Siberia. A great ball of flame brighter than the Sun was seen leaping from a forested region near the Tunguska River. The sight of the fire was followed by the sound of an explosion powerful enough to level trees within approximately 50 km. Earth tremors were recorded on seismographs throughout Europe, yet no large crater was formed, only many small ones. The most plausible explanation for the event is that a small comet (possibly part of comet Encke) or a large, fragile, stony meteorite struck the Earth, dissipated its kinetic energy on the forest and ground, and completely vaporized.

Three classes of meteorites have been established based on their chemical and metallurgical properties. The first group, known as stony meteorites, is composed primarily of silicates of iron, magnesium, aluminum, and other metals. These generally have a relatively smooth, brown or grayish, fused crust indented with pits and cavities. Buried inside all but a small fraction of them are small pieces of glassy minerals, called chondrules, which apparently formed from molten droplets, presumably during the formation of the Solar System.

One subgroup of the stones is the carbonaceous chondrites, which contain large amounts of carbon, water, and other volatiles that would have been driven off with the slightest heating above about 500 K. Therefore, these are the most primeval samples of matter from the early Solar

System that we have. They are doubly interesting because they contain organic compounds, such as hydrocarbons, amino acids, and lipids. These biologically important compounds evidently formed in the primordial solar nebula without the assistance of living organisms.

The second group, known as stony-iron meteorites, is a mix of stone and iron. Their brownish crust sometimes contains pockets of the yellow mineral olivine. Inside the meteorite the iron may have a vein-like or globular structure.

The third category, iron meteorites, is almost exclusively composed of iron, with some nickel. These meteorites are easily identified by their characteristic pitted, brownish exterior and high density. Cut, etched, and polished, they usually have a peculiar crystalline pattern unlike any in terrestrial iron. They show evidence of melting and signs of other heating and cooling processes.

Stones are the most brittle kind of meteorite, and they are far more fragile than irons. Even though most falls are stones, more of the recovered meteorites are irons because they are relatively easy to identify and they resist weathering.

Those meteorites which have been dated by their natural radioactivity average tens of millions of years for stones and 600 million years for irons. These are their ages only since the breakup of the larger mass of which they were probably a part. The most ancient specimens are about 4.6 billion years old, the same age as Earth. The chemical and mineralogical sequences in the different classes of meteorites indicate that they share the same heritage as that of the rest of the Solar System.

The origin of an individual meteorite obviously varies widely. Some meteoroids are primarily debris from asteroids and comets, in which case they formed by accretion in the early Solar System. However, some are ejected bits and pieces of matter from the Moon and Mars originating during the intense cratering period early in the Solar System's history or later. Another possibility is that meteorites may be descended from a few chemically differentiated asteroids that were whittled down by repeated collisions early in the

Solar System's history. In such a case, stony meteorites come from the original crusts, the stony-irons from the intermediate parts, and the irons from the core. Regardless of our ability to understand their origins, it is evident that asteroids and meteorites are representatives of the unused building material from which the Terrestrial planets formed at the birth of the Solar System.

9.4. INTERPLANETARY MEDIUM

The space between the planets is a vacuum by terrestrial standards, but it is not devoid of matter, since some gas and small, solid dust particles exist there. The particulate part of the interplanetary medium, called interplanetary dust, consists of particles blown out from the Sun's atmosphere by the solar wind, micrometeoric debris scattered by comets, and perhaps some granular powder strewn about by asteroid and meteoroid collisions. Interplanetary dust has even formed dust rings about the Sun analogous to Saturn's rings.

We have learned about interplanetary

dust from several sources. One is the zodiacal light, which is most easily observed in our Northern Hemisphere in spring after sundown in the west and in fall before dawn in the east. It appears as a faint pyramidal band of light tapering upward from the horizon along the line of the ecliptic. The spectrum of zodiacal light is a faint replica of the solar spectrum; it is produced by small particles lying in the plane of the planets' orbits that scatter solar photons in our direction.

Most direct evidence of interplanetary dust comes to us from spacecraft experiments. Electronic sensors on the skin of the spacecraft are arranged to count small dust particles as they strike the surface. From the numbers of impacts, it is estimated that the average spacing between interplanetary dust particles is many meters. The total mass of dust particles is estimated to be about 10^{20}g, or about one hundred-millionth of Earth's mass.

Most interplanetary matter is in the form of an ionized gas that comprises the solar wind. It consists of an almost continuous stream of particles, mostly protons and electrons, flowing out from the Sun's corona. As the solar wind moves forward, it forms an expanding spiral pattern due to the Sun's rotation, and its velocity increases until, several solar radii from the Sun, it equals the speed of sound in the plasma. Its velocity continues to increase as it flows outward, much as rocket gases are accelerated to supersonic velocities in a rocket nozzle. Near Earth the solar wind reaches a velocity of about 400 km/s. Beyond Earth its speed remains very nearly constant.

At Earth's distance the wind's density is down to about five protons and five electrons per cubic centimeter on average, but it can rise on occasion to 100 particles/cm^3. Compare that with the number of molecules in your room, about 10^{19} particles/cm3. The temperature of the wind particles is about 200,000 K in the vicinity of the Earth. This is less than their approximately 1 million K temperature when they were in the inner parts of the Sun's corona. The density is so very low, however, that the wind transfers no appreciable quantity of heat to the Earth.

The solar wind rushes past Terrestrial planets and flows out to the outer Solar System, the realm dominated by four giant planets, the Jovian planets.

9.5. SATELLITES OF JOVIAN PLANETS

A total of 63 satellites are known to orbit Jupiter. Among the four largest, discovered by Galileo in 1610, Io and Europa are about the size of our Moon, whereas Ganymede and Callisto are about the size of Mercury. The four Galilean satellites, Amalthea, and the three discovered by Voyager orbit within Jupiter's magnetosphere. As they orbit Jupiter, the Galilean satellites and Amalthea keep the same face toward the planet as the Moon does with respect to Earth. The Galilean satellites, Amalthea, and the innermost small satellites form what can be called a regular satellite system in that they orbit in nearly circular orbits in Jupiter's equatorial plane and revolve in the same sense as Jupiter rotates. The outer moons are much smaller than the Galilean satellites and move in irregular orbits inclined at varying angles to Jupiter's equatorial plane. They form an irregular satellite system in many respects.

Before the Pioneer and Voyager spacecraft sailed past Jupiter and Saturn, astronomers were not generally aware that there were any similarities between the Galilean satellites and the Terrestrial planets. Even though there are compositional and internal structural differences, the Galilean satellites have and continue to evolve under processes similar to those operating in the Terrestrial planets. Astronomers now give the Galilean satellites a great deal of attention. In order of distance from Jupiter, their mean densities in g/cm³ are as follows: Io, 3.55; Europa, 3.04; Ganymede, 1.93; and Callisto, 1.81. Hence Io and Europa, with size, density, and mass comparable to the Moon, probably have a rocky, silicate-rich composition and structure similar to the Moon, whereas Ganymede and Callisto are lighter and made of a mixture of rocky and icy matter.

There are fascinating differences in the surface appearances of the Galilean satellites. Europa, Ganymede, and Callisto have icy crusts with surfaces that are covered by ice, or a mixture of ice and rocks, many kilometers thick, whereas Io is quite different and a most exotic body.

Saturn has some 62 satellites. Like Jupiter, Saturn has a regular satellite system of bodies moving in near circular orbits in the planet's equatorial plane. Their direction of motion is the same as Saturn's direction of rotation. Iapetus and Phoebe are part of an irregular satellite system with orbits inclined by several tens of degrees to Saturn's equatorial plane. Phoebe's motion is retrograde, or opposite to Saturn's direction of rotation.

What causes the irregular satellite systems of the Jovian planets? They may be captured bodies, such as pristine comets or icy-composition asteroids.

Finally, whereas Jupiter has four planet-size satellites, that is the Galilean satellites, Saturn has only one, Titan, that is that large.

Io surprised Voyager scientists by its surface appearance, which is a collage of mottled yellows, reds, and blackish browns. Passing very close to Io, Voyager 1 was able to resolve features smaller than 1 km. The satellite has a thin, patchy atmosphere, with sulfur dioxide as its primary constituent. The greatest excitement about Io is the positive identification of active volcanoes on its surface. The first such image with an eruption was recorded in March of 1979 and shows an eruption occurring on the limb, with material being thrown up to altitudes of about 150 km at velocities of about 1 km/s. Such high speeds suggest that these are not Earth-like volcanic eruptions, since the latter seldom exceed 0.1 km/s. At least eight active volcanoes were identified in the Voyager 1 photographs, and seven were captured erupting four months later when Voyager 2 arrived.

Launched in 1989 and arriving at Io in 1995, the Galileo spacecraft was able to completely map the surface of Io. It proved Io to be the most volcanically active celestial object in the Solar System. Its photographs were nothing less than spectacular.

The bright array of colors of the surface is due to sulfur compounds from volcanic activity. As you may remember from a past chemistry course, sulfur is normally a bright yellow. If heated and cooled quite rapidly, however, sulfur can take on a range of colors from orange and red to black. Related to the discovery of volcanic activity is the fact that no impact craters, such as on the Moon, can been seen on Io's surface. There are also hot spots, up to 500 K, on the surface, whose typical temperature is 60 to 120 K. Finally, some 200 calderas, both with and without lava flows, dot the surface; Earth has only 15 or so. Thus the surface is active enough that impact craters are either eroded away or filled in by volcanic debris in time periods as short as 1 million years. Io must possess the youngest surface of any Solar System body we have examined, and it is the only body besides Earth to show significant volcanic activity, which is actually greater than that of Earth.

The probable reason for the extensive volcanic activity is heating by tidal friction. Moving in an elliptic orbit, Io slightly shifts in and out in Jupiter's powerful gravitational field, with the result that it is flexed rapidly by tidal forces that release an enormous amount of frictional heat in its interior. This thermal energy, as it works its way to the surface, powers the volcanism. Io's volcanic activity may be akin more to the eruption of a hot water geyser on Earth, such as Old Faithful in Yellowstone Park, than to a volcanic eruption, such as Mount St. Helens.

In stark contrast to Io is the next Galilean satellite out from Jupiter, Europa, This satellite is Moon-like in size, density, and mass but has a much higher reflectivity than the Moon, which indicates an ice-rich surface. There is very little relief to the surface, which has been likened to a frozen ocean of water 100 km deep. The surface color is a lightly orange-hued off-white and has the least contrast of the four Galilean satellites. There appears to be a total lack of craters larger than about 50 to 100 km across and very few smaller ones. The low number of craters is indicative of a young surface, something less than 3 billion years of age.

What makes Europa the enigma that it is are the vast numbers of crisscrossing light and dark markings. These features are tens of kilometers wide, and some are thousands of kilometers long. In fact, some appear to extend halfway around the satellite and may be up to 3500 km in length if they are truly continuous. These stripes are thought to be cracks in the thick icy surface layers, caused by tidal flexing by Jupiter, into which darker material has been forced.

Ganymede, which has a smaller mean density than Europa, as is also the case for Callisto, must possess a much thicker layer of water ice surrounding its rocky interior than the 100-km surface layer on Europa. Estimates suggest that both Ganymede and Callisto possess 1000-km thick mantles of water ice.

Voyager photographs of Ganymede and Callisto reveal heavily cratered surfaces, like the ancient surface of the Moon, with the exception that these are craters made in an icy surface. Ganymede has two fairly distinct types of terrain. A dark terrain appears to be the oldest part of the surface, in that it is heavily cratered, whereas a lighter-colored, heavily grooved terrain possesses a smaller number of craters and is therefore younger. On Terrestrial planets, young craters are bright, often having a bright ray system of ejected material, while old craters are dark and lack ray systems. In addition to bright, rayed craters on

Ganymede, there are also some dark-rayed craters that are relatively young; these craters suggest that ice-dominated soils can behave differently from the silicate-rich soils of Terrestrial planets.

Lighter-colored terrain on Ganymede displays a number of features reminiscent of tectonic activity on Terrestrial planets, which further supports the contention that it is the youngest part of the surface. Dominating the light regions is a system of grooves that is very unlike the dark stripes on Europa. The grooves are parallel sets of ridges and troughs whose widths are up to tens of kilometers, and they may be a few hundred meters deep. Forming bands or sets of grooves, they wander for thousands of kilometers across Ganymede's surface to intersect in intricate patterns. As mentioned, the grooves have craters superimposed on them, so they are not very recent additions to the surface. The offset of grooves across what appear to be fault lines suggests that the breakup of dark crustal blocks is responsible for the grooves. Thus a global scale of tectonic activity may have existed on Ganymede some time during the first one and a half billion years of its existence.

The outermost of the Galilean satellites is Callisto. It is a bit smaller and a little less dense than Ganymede, and it too probably has an ice-rock composition. Callisto has about 10 times as many craters as Ganymede on its bitterly cold surface. Daytime temperatures reaching only -118°C and nighttime temperatures down to -193°C. In fact, craters dominate the entire surface and stand nearly shoulder to shoulder. Callisto is unique in having no plains or regions where later processes have obliterated the craters. It also appears to lack fractures in its crust, which Ganymede has.

Callisto has several large, circular impact basins that are surrounded by an almost concentric sequence of rings. The rings are raised, separated by 50 to 200 km, and possessing diameters up to 3000 km. For many features on Callisto there does not appear to be a significant difference in elevation, which is somewhat puzzling. This characteristic may be an indication of a relatively weak surface material that, because of icy composition, is unable to support much vertical relief.

Titan is another of the planet-size icy bodies of the outer Solar System like the Galilean satellites of Jupiter. Estimates are that it must be 52 percent rocky and 48 percent icy materials in order to have a mean density of almost 2 g/cm^3. It is unique, and thus highly intriguing, in that it is the only known satellite to possess a substantial atmosphere. Io's atmosphere is rarefied and very spotty in its density. Above a unit of surface, Titan's atmosphere contains more gas by a factor of 500 than does the atmosphere of Mars.

Titan and Earth are the only bodies in the Solar System with atmospheres dominated by molecular nitrogen. Estimates for Titan lie between 82 and 94 percent for nitrogen, about 12 percent for argon, and a few percent for methane, which was first discovered in 1944. In addition to these primary constituents, there is a smattering of molecular hydrogen that is escaping from the satellite and hydrocarbon molecules.

In the very complicated atmospheric chemistry of Titan, methane clouds form some 3 km above the surface and smog particles form from about 200 km down to the surface. The smog is so thick that nowhere are there holes through which the surface could be seen by the Voyager cameras. Thus the satellite has a rather bland reddish-brown color, with the only markings a polar hood and a change in reflectivity at the equator.

In addition, the low surface temperature has prompted speculation that methane, or natural gas, might exist in gaseous, liquid, and solid forms at the same time near the surface of Titan. Thus there could be oceans of liquid methane in warmer regions and methane polar ice caps. Whether or not continents of icy soil and rocklike ice divide the methane oceans is pure speculation. However, methane ice clouds could exist in the low atmosphere, from which methane rain falls on occasion. Methane could apparently play the same role on Titan that water plays on Earth, and thus they may be unique in the Solar System in having substantial amounts of liquid covering their surfaces.

We have learned much about Titan from the Cassini spacecraft and its Huygen's probe which landed on the surface of the only Solar System moon with a dense atmosphere. In fact, atmospheric pressure on the surface of Titan is equivalent to that on the surface of the Earth.

The intermediate-size satellites in order of distance from Saturn are Tethys, Dione, Rhea, and Iapetus. Rhea and Iapetus are the second and third largest of Saturn's satellites, with nearly identical radii of about 750 km. The fourth and fifth largest are Dione and Tethys, with radii of about 550 km. All four have about the same mean density of between 1 and 1.5 g/cm^3, and thus they are all presumably icy conglomerates. Iapetus is part of the irregular satellite system, whereas the others are part of the regular system. Iapetus is a strange body in that half its surface is bright and the other half nearly black. The dark hemisphere is the one that appears to always face forward as the satellite orbits Saturn for reasons of which we are not sure.

Rhea and Dione possess heavily cratered leading hemispheres, whereas their trailing hemispheres are covered by a network of strange wispy markings. These wisps may be troughs and valleys in the icy surface, or they may be fresh deposits of frozen water formed by outgassings from the satellites' interiors. Tethys, the closest to Saturn of the intermediate size satellites, is completely different: Its surface is heavily cratered and shows only small variations in brightness globally. Thus it does not have the hemispheric pattern of Rhea and Dione.

The Voyager spacecraft have given us a glimpse of new worlds in the satellites of Jupiter and Saturn that could not have been imagined from our Earth-based studies. As is so often the case, the glimpse has raised more questions than it answers. Scientists are hopeful that several new ventures to the outer Solar System will someday be approved.

Uranus has five satellites that are visible in ground-based telescopes. These are Ariel, Umbriel, Titania, Oberon, and Miranda. After Voyager 2's visit, ten satellites were added to the list of known Uranian satellites. Now there are about 27 known satellites of Uranus. Two of the satellites orbit close to the planet's largest and outermost ring. Eight satellites follow nearly circular orbits that lie inside the orbit of Miranda, the innermost of the five originally known satellites.

Ariel, Umbriel, Titania, and Oberon are quite similar in size and approximately the size of the Saturnian satellites Tethys, Dione, and Rhea. Miranda is considerably smaller than the other large Uranian satellites. Spectra of reflected sunlight from the surfaces of the larger Uranian satellites suggest the presence of water ice on their surfaces. And since their densities are between 1.3 and 1.6 g/cm^3, it is likely that the Uranian satellites are similar to the Saturnian satellites being composed of about half silicate rock and half icy materials, primarily water ice. Judging from the Voyager 2 pictures of the five largest satellites, all, and especially Ariel and Miranda, seem to have been geologically active early in their history. Although all show evidence of having undergone an early and intense impact cratering period as do other bodies in the Solar System, several of their surfaces bear the dramatic evidence of global tectonics in the form of rift valleys and suggestions of lava flooding.

Triton, Neptune's major satellite, is one of the larger satellites in the Solar System.

Neptune is known to have some 13 other satellites, all much smaller than Triton. Triton itself was discovered only a month after the planet itself was discovered in 1846. Mass and radius determinations for Triton were quite difficult to make and were, consequently, very uncertain until Voyager 2 arrived in 1989. The diameter of Triton is 2700 km and its density is 2 g/cm³. Triton is primarily an icy body. Triton is unusual in a number of respects. It is the only large moon of a planet that orbits its planet in the opposite direction of the planet's rotation. This, and its icy nature, leads some scientists to believe that Triton may be a captured Kuiper Belt Object. The biggest surprise about Triton was the discovery of active geysers on its surface.

9.6. COMETS: ICY MESSENGERS FROM THE PAST

Comets are among the most spectacular objects in the Solar System and appear unexpectedly in all parts of the sky. They are often discovered accidentally in photographs taken by professional astronomers for other purposes or by amateur astronomers methodically searching for them. As an honor, comets are named after their discoverers. Only once every other year or so does a comet become bright enough to be seen with the naked eye. A spectacularly bright comet appears about once or twice each decade. The usual telescopic comet appears as a small hazy object with a roundish nebulosity, called a coma, or head, and occasionally a short tail.

Bright naked-eye comets, however, possess a more interesting structure, in that they have a large, teardrop-shaped coma surrounding a small bright nucleus and a well-formed tail pointing away from the Sun. While observing bright comets from night-to-night, it is apparent that their coma and tail undergo changes in appearance, sometimes making observable changes in matters of hours. The changing appearance results from the flow of matter from the nucleus into the coma and from there out into the tail.

These observations, along with satellite studies, show that comets have four principal parts: a nucleus or the principal mass of the comet, a nebulous coma surrounding the nucleus, an invisible hydrogen cloud enclosing the coma, and a tail or two. The size of the coma may vary from tens of thousands of kilometers to well over a million kilometers. Surrounding the coma, an immense hydrogen cloud spans millions of kilometers. The comet's mass is concentrated in the nucleus, which is the actual comet and which appears to be on the order of many kilometers in diameter. Well-developed tails usually form when the comet is within Earth's orbit and can be millions to hundreds of millions of kilometers long. Very rarely will the tail be longer than Earth's distance from the Sun. There can be two distinct types of tails, with both often present. One is a yellowish sweeping arc of dust, and the other is a long, straight, bluish tail of plasma, or stream of ions, charged particles.

The first positive evidence that comets are extraterrestrial objects was found by the sixteenth-century astronomer Tycho Brahe. He tried to find the parallactic displacement of the comet of 1577 relative to the background stars by comparing measurements from his observatory and with those from other European centers. He decided that the object was more distant than the Moon. A century later, Isaac Newton and Edmund Halley demonstrated that comets are members of the Solar System and move in elliptic orbits under the gravitational attraction of the Sun. For example, old records reveal that the most famous comet, Halley's Comet, has been observed at every return since 239 B.C. Its appearance in A.D 1066 is recorded in the historic Bayeux tapestry.

Comets fall into two groups, long-period and short-period comets, depending on the period of orbital revolution around the Sun. Of the 600 or so for which information exists, about 500 are long-period comets with periods varying from thousands to millions of years, and approximately 100 are short-period comets with periods of less than 200 years.

The long-period comets, which astronomers believe are the vast majority of all comets, travel in highly elongated ellipses inclined at all angles to the ecliptic plane. The brighter members are usually among the most magnificent comets. Some of these comets are approaching the Sun for perhaps the first time. Several comets in the long-period category have grazed the outer parts of the Sun. Frequently, forces produced by the Sun break them apart during this close encounter, and the fragments travel on as independent comets along orbits nearly identical with the parent's orbit.

By contrast, the short-period comets, orbits the Sun at small or moderate angles of inclination to the ecliptic plane, nearly all in the same sense as the planets. Roughly, every third or fourth new comet discovered has a short period, ranging from 3.3 to 200 years. Approximately 50% have their aphelion, which is the point in the orbit most distant from the Sun, somewhere near Jupiter. Following up this suggestive face, one finds that, when their orbital history is traced back mathematically, the short-period comets initially moved in long-period eccentric orbits, bringing them on one critical occasion into a chance encounter with Jupiter. Thus all comets are initially long-period ones. Apparently, the great planet's gravitational attraction has so modified their orbits that they now form a Jovian family of short-period comets.

Astronomers know that comets overall are rather flimsy structures of low density from the following evidence: First, they cannot be observed on the solar disk when they pass in front of the Sun; secondly, we can see stars through the tail and the outer portions of the coma; third, there are the changes from night to night in brightness and size in the coma that are even more pronounced in the tail; and fourth, they are perturbed by the solar wind and by tidal, gravitational, or other disruptive forces.

Today most astronomers agree that comets are just large "dirty icebergs," that is, icy conglomerates composed mostly of frozen water with some carbon dioxide and other ices impregnated with small pieces of particulate matter and fine dust. This conglomerate is the nucleus of the comet, where the bulk of the mass is located.

Being a flimsy structure as a comet approaches the Sun, the surface of the comet is vaporized by solar radiation releasing matter to form the coma. The solar ultraviolet radiation breaks complex molecules down into simpler molecules of hydrogen, carbon, oxygen, nitrogen and sulfur. These molecules are identified by the bright bands they produce in the emission spectrum of the comet. Emission lines of gaseous sodium are also present in the spectrum, along with some lines of iron, magnesium, and silicon when the comet comes very near the Sun. The emission lines and bands are superimposed on the weak background spectrum of sunlight reflected from dusty material.

The nuclei of short-period comets have lifetimes of a few millennia, since they lose about one percent of their mass on each perihelion passage. They eventually evaporate away their basic constituents of gas and dust, leaving a rocky remnant that should be indistinguishable from an asteroid or meteoric debris in appearance.

The comet's head plowing through the onrushing solar wind creates a bow shock wave. High-energy electrons in the solar wind ionize the molecular gases in the coma. Ultraviolet photons from the Sun probably dissociate the hydroxyl radical, releasing the hydrogen to form a huge hydrogen cloud around the coma. Chaotic magnetic fields in the solar wind sweep the charged molecules away from the coma at high speeds, forming the narrow, bluish ion tail. What causes the wide, yellowish, curved tail? The Sun's electromagnetic radiation can push the dusty material flowing from the coma at different velocities away from it at comparatively low speeds.

The Dutch astronomer, Jan Oort, studying how cometary orbits are distributed, suggested in 1950 that a cloud of comets, the Oort Cloud, of not more than a few Earth masses surrounds the Sun at an average distance of 50,000 AU. For comparison, the nearest stars are over 300,000 AU from the Sun. Detached from this great reservoir, the actual number may be in the tens of millions, by perturbations from nearby stars, a few begin to orbit the Sun as long-period comets. Around 100,000 comets might have come close enough to the Sun to be observable.

From their apparent structure and composition, it seems probable that comets are primeval material, basically unchanged since the origin of the Solar System some 4.6 billion years ago. Comets appear to be a link between the Solar System and the interstellar medium. Many complex molecules have been discovered in dark clouds, or dark nebulae, in the interstellar medium. When frozen, could these form icy structures, similar to a portion of comets?

Chapter 10
Formation of the Solar System

Now that we have acquainted ourselves with the nature of the Solar System, with the exception of the Sun, it seems a reasonable step to inquire into what astronomers think that they understand as to the origin of the Solar System and the reason for the diversity of its members. In the preceding chapters we have touched at several points on the origin and evolution of the Solar System. It is now time to bring these scattered ideas together into a coherent theory of that momentous event which occurred some 4.6 billion years ago.

10.1. ARCHITECTURE OF THE SOLAR SYSTEM

As to the question of how the Solar System began, there are theories that have been proposed to explain that event. One such category of theories is those theories that invoke an accidental catastrophic event, such as the near collision between the Sun and a star, while the other category of theories is those theories that involve a natural, non-catastrophic event, such as might occur in conjunction with the birth of any star. Here we need only say that we know that stars are born, live out their lives, and die, just as the things of Earth are not eternal. A historical approach to the explanations of planetary genesis is a good beginning, but first we should summarize the major characteristics of the Solar System for which any theory of origin should account.

A sequence of natural forces evidently created and shaped the Solar System somewhat along the lines revealed by the following clues, which suggest that the design most likely possessed a continuity in its processes and did not materialize through a sequence of unrelated, random events. The overarching observations include:

- the planets are isolated from each other without bunching, and they are placed at orderly intervals;
- the planets' orbits are nearly circular, except for those of Mercury;
- the orbits are nearly in the same plane; Mercury again most

exceptional;

• the planets and asteroids revolve around the Sun in the same direction that the Sun rotates (from west to east);

• the planets also rotate around their axes from west to east, except for Venus and Uranus;

• a planet's system of satellites can be divided into either a regular system with direct orbits approximately in the planet's equatorial plane or an irregular system with irregular orbits inclined at various angles;

• the Terrestrial planets have high mean densities and relatively thin or no atmospheres, rotate slowly, and possess few or no satellites, points that are undoubtedly related to their size and closeness to the Sun;

• the giant planets have low mean densities, relatively thick atmospheres, and many satellites, and they rotate rapidly, also related to their masses and distances from the Sun;

• the terrestrial planets are rocky bodies that are poor in hydrogen;

- the Jovian planets are large, low-density planets that are fluid-like and are rich in hydrogen; and,
- most of the outer planets' satellites, comets, and Pluto are icy bodies.

10.2. NEBULAR HYPOTHESIS

The German philosopher Immanuel Kant speculated in the middle of the eighteenth century that the Solar System had been formed out of a huge rotating gaseous nebula slowly contracting and condensing. A nebula is a large cloud of gas and dust particles, held together by the mutual gravitational attraction of the particles composing it. Such nebulae, as we see them elsewhere in our galaxy, are immensely larger than the Solar System. Pierre Laplace (1749-1827), a French mathematician and astronomer, expanded the idea in 1796, and it became known as the nebular hypothesis.

Laplace theorized that as the large, slowly rotating solar nebula of hot gaseous matter contracted; it rotated faster and faster, flattening into an equatorial ring. The physical principles involved here are the action of gravity and the conservation of angular momentum, which requires a spinning body to rotate faster as it shrinks. The angular momentum of a rotating body, a measure of its quantity of rotation, remains constant unless energy is taken out of rotation and put into some other form. If the radius of the body decreases, the rotational velocity must increase to compensate for the reduced radius; this is what we observe when we see a spinning ice skater rotate faster as he or she brings his or her outstretched arms closer to their body, his or her angular momentum is thus conserved.

Laplace supposed that when the centrifugal force acting on the outer rotating edge of the solar nebula exceeded the inward gravitational force of the nebular mass, a ring of gaseous matter was split off, eventually coalescing into a planet. The splitting process repeated itself, making concentric rings that formed into planets, whereas the central portion condensed to become the Sun.

The theory has two major defects. First, whereas 99 percent of the Solar System mass resides in the Sun, 99 percent of the angular momentum of the system resides in the planets' orbital and rotational motions. We might guess intuitively that the distribution of angular momentum in the forming Solar System ought to roughly match the distribution of mass; the central mass could not have transferred this much momentum to the planets. Second, a hot gaseous ring of the type postulated would disperse into space and not pull itself together gravitationally to form a planet.

10.3. ENCOUNTER THEORIES

At the beginning of the 20th century, attempts to reconcile the nebular hypothesis with physical principles were temporarily abandoned. A different approach, the so-called encounter theory, which had been conceived in 1745 by the French naturalist Georges Buffon (1707-1788) when he proposed that material ripped off from the Sun by collision with a comet had condensed into the planets, was taken by the American geologist Thomas Chamberlin (1843-1928) and the American astronomer Forest Moulton (1872-1852). They suggested that giant eruptions were pulled off the Sun by the gravitational attraction of a passing star.

Somewhat later another geologist-astronomer pair in England, Harold Jeffreys (1891-1989) and James Jeans (1877-1946), theorized that a cigar-shaped gaseous filament was pulled from the Sun by the sideswiping action of a passing star. The middle section condensed into the Jovian planets, and the ends condensed into the smaller planets.

The encounter theory accounts for the common direction of the planets' orbital motion and the Sun's rotation as well as for the planets' nearly circular and coplanar orbits. In either version, however, this theory has serious failings in that solar matter, whether pulled or ejected, could not have acquired sufficient angular momentum nor could hot gas have condensed into planets. Besides, the probability of a near encounter in our region of the Galaxy is vanishingly small, less than one in many millions.

10.4. PROTOPLANET THEORY, THE SOLAR NEBULA

By mid-century, astronomers once more turned their attention to possible improvements in the nebular hypothesis. A new factor was introduced in the form of the existence in the cool gaseous nebula of a small amount of dust, providing nuclei for the condensation of gas particles into larger aggregates that could accrete and solidify into the embryo planets. The existence of dust particles in the interstellar gas clouds out of which stars are formed was accepted in the 1930s thanks to Robert Julius Trumpler (1886-1956). This modern version of the nebular hypothesis is sometimes simply called the protoplanet hypothesis, and it owes much of its recent revival to the power and scope of computer analysis. It was first formulated independently by Carl von Weizsacker (1912-2007) and by Gerard Kuiper (1905-1973) in 1945 and then extended and modified over the years by others.

The hypothesis begins with a fragment

separating from an interstellar cloud composed mainly of hydrogen and helium, with trace amounts of the other elements. With other fragments of the interstellar cloud presumably following a similar evolution, its central region, being somewhat more dense, collapsed more rapidly than its outlying parts. This formed the central portion of the solar nebula, whose outer portion contained a thin disk of solids within a thicker disk of gases. The original interstellar cloud must have been rotating, and as it fragmented, rotation was imparted to each fragment. Thus as the solar nebula contracted, it rotated more rapidly, conserving angular momentum.

The solar nebula grew by accretion as material continued to fall inward from its surroundings. Large-scale turbulence from gravitational instabilities ruptured the thin disk into eddies, each containing many small particles. These particles gradually built up into larger bodies by some combination of adhesive forces. Repeated encounters among them resulted in the accretion of literally billions of still larger asteroid-sized aggregates called planetesimals, which orbited the center of the solar nebula. Mutual gravitational attraction led to further encounters and gradual coalescence into many roughly Moon-size bodies, which in turn coalesced to form the planets.

Planetesimals must have differed in chemical composition, depending primarily on their initial distance from the Sun as it formed. That is, as the central portion of the solar nebula contracted, the temperature rose to around 2000 K, hot enough to vaporize all compounds in the dust except the "high-temperature" metallic and silicate minerals in the inner portion of the disk, while the outer disk remained relatively cool. Planets that formed close to the young Sun, such as the Terrestrial planets, would be expected to contain less of the volatile icy and gaseous materials and thus be richer in the rocky materials.

10.5. FORMATION OF THE PLANETS

During and following the formation of the Terrestrial planets, there was a catastrophic bombardment by the remaining rocky planetesimals that cratered the surfaces of these planets. The impacting material, coupled with intense radioactivity and subsequent gravitational concentration, produced sufficient heat to melt and chemically differentiate the planets into their presently layered structure, that is, core, mantle, and crust. The atmospheres of the Terrestrial planets were formed during this process and afterward by outgassing from impacting material and from the hot interiors of the planets.

In the asteroid belt between Mars and Jupiter, the temperature of the solar nebula was lower so that carbon- and water-rich minerals could coalesce in the forming planetesimals. From about Jupiter outward, temperatures were even lower, so that huge amounts of frozen water could accumulate with the rocky material in the planetesimals. At still colder temperatures, other ices would have formed, such as ammonia and methane, giving those distant planetesimals a mixed composition of water, ammonia, and methane ice impregnated with a small amount of rocky matter.

Within a relatively short time after contraction of the solar nebula began the young Earth had collected most of the matter that composes it today. Matter attracted by the growing Earth collided with it, giving up its kinetic energy as heat. This energy, along with the energy resulting from Earth's gravitational contraction and emissions by radioactive nuclei, heated Earth's interior.

In a few tens of millions of years, the Earth became molten; chemical differentiation followed. The heaviest elements, iron in particular, separated from the lighter elements, such as oxygen and silicon (primarily in the form of silicates and oxides of iron and magnesium) and sank toward the center. The silicates and oxides rose to form the mantle surrounding an iron-rich core. The lightest materials rose to the top and solidified as the crust.

About 4.0 to 4.5 billion years ago, Earth as a whole was cooling even though volcanic activity on the surface was intense. We believe that during this period an atmosphere of whose composition we are not certain was formed, probably from the gases carbon dioxide, carbon monoxide, nitrogen, water vapor, and possibly some hydrogen sulfide and hydrogen. As Earth cooled, these gases escaped from the interior during volcanic activity and water condensed, forming the oceans. But what of our Moon during this period?

Even with all our new information from the Apollo program, the Moon's origin, like that of the Solar System itself, is still shrouded in mystery. That is not to say that having astronauts trudging over the surface of the Moon was of no benefit. Quite the contrary, the information derived from the astronaut landings on the lunar surface were of inestimable value. However, we now know that missing from the Apollo samples are those oldest of rocks, pristine remnants form our satellite's formation 4.6 billion years ago, that conceivably could settle the question of the Moon's origin forever. As it stands now, no proposal for a beginning for the Earth-Moon system is without some objections.

Prior to the Apollo program, there were three concepts that dominated lunar-origin theories. The earliest concept, the fission theory, was that Earth was spinning rapidly and flattened to a dumbbell shape perhaps because of movement in the Earth's molten core. The smaller end of the dumbbell tore away from the primitive Earth to become the Moon, separating ever more from the Earth because of tidal forces. The major objection to this theory is that the primitive Earth could not have spun rapidly enough to promote fission through rotational instability. A second objection is that, compared with Earth rocks, lunar rocks have slightly greater proportions of those elements that are difficult to vaporize and slightly less of the easily-vaporized elements. This suggests that the Moon formed from material somewhat hotter than that from which the Earth formed.

The second concept is that the Earth-Moon system formed by accretion from chemically related primordial planetesimals condensing out of the gas and dust of the solar nebula; this is the condensation or co-accretion theory. The essence of this idea is that the Earth and Moon are actually a double planet system with colliding debris trapped in orbit around the growing Earth accreting to form the Moon. A fact which supports this theory is that the two bodies are of comparable ages, about 4.6 billion years. Also, since chemical analysis of lunar rock samples shows some chemical disparities between the Earth and the Moon, the Earth and the Moon may or may not have evolved from the same parent material. At any rate, if it was the same parent material for both bodies, then some aspect of the formation process permitted the chemical disparities to arise.

Finally, a third concept is that a proto-Moon originally was moving in a highly eccentric orbit around the Sun. As it approached Earth almost on a collision course, it was disrupted by strong tidal forces, and most or some portion of the fragmented body became Earth's satellite, the capture theory. The problem with this idea is that capture is not easy to accomplish. Although it apparently can happen, some means must occur to remove some of the kinetic energy of the captured body so that it moves from a solar to an Earth orbit. Just passing by a more massive body does not automatically lead to capture.

In the post-Apollo era, the Apollo samples impose strict restrictions on the three classical lunar-origin theories such that all three are found wanting. It is unfortunate that the Apollo samples in themselves do not provide the answer. However, such is not the case and these samples can only serve to provide boundaries within which any seriously considered theory is constrained to lie. A post-Apollo theory, which is called a collisional ejection theory, is to assume that a Mars-sized planet struck the primitive Earth in part coalescing with Earth and in part ejecting a cloud of material to orbit Earth. This cloud of hot material eventually cools over many centuries and forms the Moon. The chief advantages of this theory are its ability to account for the near similarities, but distinct differences, in the chemical composition of the Earth and Moon. Its disadvantages are mostly dynamical problems in that the lunar orbital plane is not coincident with the equatorial plane of Earth. Thus the collisional ejection theory, like its classical predecessors, is unable to provide all the answers to the long-studied mystery of the origin of the Earth-Moon system.

Within the outer, cooler regions of the

solar nebula, the icy planetesimals collided, building larger bodies of ice and rock. As these bodies grew to a mass a few times that of the Earth, then they drew in more hydrogen and helium from the surrounding interplanetary gas. Naturally, capture and retention of gas were easier far from the Sun, where the temperature was lower. Because of their great masses, they have kept very nearly the same relative proportion of hydrogen and helium to the heavier elements as the Sun and the interstellar medium have. This is the most likely mode of formation for Jupiter and Saturn and why they are hydrogen-rich bodies. Uranus and Neptune were simply never massive enough to accrete hydrogen and helium to the extent that Jupiter and Saturn did. Thus carbon, nitrogen, oxygen, silicon, and iron dominate their compositions. The comets are probably a fossil relic of the primordial icy planetesimals that existed in the outermost regions of the solar nebula.

In the regular satellites of Jupiter and Saturn we probably have a miniature version of planet formation. Contraction by Jupiter and Saturn at the time of their formation released a great deal of gravitational potential energy, heating them significantly. Jupiter was some 10 times brighter than at present, so that the contracting planet raised the temperature of matter close to it, accounting for why the two inner Galilean satellites are rocky bodies and the two distant ones are primarily icy bodies. This parallels the decline in mean density with distance from the Sun found in the planets themselves.

When the protoplanets were all formed, the solar nebula's central bulge rapidly collapsed into the protosun. Continued contraction raised its internal temperature from a few tens of thousands of degrees to several million degrees when the first stages of nuclear burning were initiated. In the last stages of formation, the Sun may have had a much more intense solar wind, which presumably blew away much of the primordial gas and dust left over from the original interstellar cloud. But this point is still pretty much of a mystery.

A weakness in the protoplanet hypothesis is that it does not provide a completely satisfactory explanation for the observed distribution of angular momentum in the Solar System. If the angular momentum of the planets could somehow be returned to the Sun, its present slow rotation, like that of stars similar to it, of 2 km/s would be increased to about 100 km/s. The primitive Sun apparently transferred most of its angular momentum to the planets as they were forming.

To explain this transfer of angular momentum, astronomers have proposed a braking action caused by what is called magnetohydrodynamic forces on the Sun as its magnetic field interacted with the ionized nebular gas in the disk. The magnetic lines of force spiraling outward from the rotating Sun into the surrounding nebula would act as a magnetic drag on the spinning Sun and serve as conduits, transferring angular momentum to the planetary disk.

There is a recent discovery of a relatively high abundance of some rare, by Earth standards, isotopes in primitive meteorites. It has been proposed that the isotopic anomalies are due to the injection of matter from a supernova explosion into the Solar System a few million years before the meteorites solidified. A supernova is the explosion of a star in the last stages of its life. Possibly the concussion from the explosion triggered the collapse of the interstellar cloud to form the solar nebula.

Chapter 11
Radiation from the Sun

The theory of the discrete nature of light began a conceptual revolution in twentieth-century physics and astrophysics. It was used by Niels Bohr to formulate a new model for the atom that can be used to understand how light is created and destroyed inside atoms in distance stars. Using what they know about the properties of electromagnetic radiation, the atom's structure, the interaction between matter and energy, and spectrum analysis, astronomers can study the Sun and stars by means of the radiation they emit. Moreover, as we develop even greater understanding of the nature of radiation and its interaction with matter, we can explore more deeply the dim sources of radiation, which are galaxies and clusters of galaxies, in the outer reaches of the cosmos, almost back to the beginning of time.

11.1. STRUCTURE OF ATOMS

One of the most perplexing problems

413

for early twentieth-century physicists was why atoms emit a discrete pattern of spectral lines, that is to say, photons with only selected wavelengths. To understand why this was a problem, let us back up a few years to Maxwell's electromagnetic radiation theory which was advanced in the latter half of the nineteenth century. Maxwell's concept of a field provided a framework whereby electricity, magnetism, light, and even gravity could be seen theoretically as related ideas through "action-at-a-distance" and energy. His concept of light as being an electromagnetic wave composed of moving electric and magnetic fields which carry energy and that such waves are emitted by fluctuations in electric currents was a bold unification of seemingly unrelated phenomena.

Experimental confirmation was provided by the German physicist Henrich Hertz (1857-1894), who undertook in 1887 to show that oscillating electric currents send out electromagnetic waves possessing all the properties of light except visibility. Although Hertz's electromagnetic waves were microwaves, his experiment was overwhelmingly successful in convincing those who doubted Maxwell's theory that it was correct. And in so doing, he confirmed that electromagnetic waves are generated by accelerating electric charges. Consequently, there developed a theory for the emission of light as being due to the oscillatory motion of electric charges located in the atoms of the radiating source.

This theory had many early successes and from it was sought an explanation of the various colors, photons of different energies, emitted by radiating sources. The discovery by Thompson in 1897 of the electron as a negatively charged constituent of atoms seemed to provide the needed accelerated charge for producing electromagnetic waves. If the electron moved in a circular orbit inside the atom, for example, then in Newtonian mechanics that is accelerated motion since the direction of motion is constantly changing. Hence the accelerating electron should emitted electromagnetic waves. But, since electromagnetic waves carry away energy, this means that the electron should lose energy and consequently spiral in an ever tighter circle about the nucleus. This tightening spiral motion should cause the emission of waves with a continuous range of wavelengths, resulting in the electron spiraling ever closer to the nucleus until its energy is gone and it is pulled into the nucleus. Rutherford was to later say that, "I was perfectly aware when I put forward the theory of the nuclear atom that according to classical theory the electron ought to fall into the nucleus...." This theory seemingly accounted for continuous spectra, but did nothing to account for the discrete nature of emission spectra. Also in this model, why did the atom sometimes emit electromagnetic waves and not at other times? The answer to the dilemma lay in the concept of the photon by Planck and Einstein, which Bohr was to incorporate with Rutherford's nuclear atom.

In 1913, by which time the structure of the atom was reasonably well understood, Bohr proposed a theory for the orbital structure of the electron in hydrogen. He envisioned a hydrogen atom as being like, in some respects, a miniature planetary system extending about a tiny nucleus composed of the one proton. Orbiting the nucleus is the single electron, like a tiny planet moving in roughly a circular orbit held by the electric force of attraction between the proton and electron rather than Newton's gravity. However, and most importantly, unlike the gravitationally bound Solar System, Bohr made the following hypotheses concerning the behavior of an electron in the hydrogen atom:

- the electron can occupy only a selected number of prescribed concentric orbits with discrete radii rather than having an unlimited and unspecified orbital distance;
- when the electron exists in anyone of its permitted orbits, electromagnetic energy is not radiated away; and,
- the electron can temporarily occupy a higher-energy orbit from which it spontaneously disappears and reappears in a lower-energy orbit giving off the energy difference as a photon of electromagnetic radiation.

In Bohr's theory, orbits representing higher levels of energy are increasingly farther from the nucleus. And, it is only when an electron changes from one of its permitted orbits, which is higher in energy to another of lower energy, that a photon of electromagnetic energy is emitted. In addition, Bohr indicated that the electron normally resides in the lowest-energy orbit, which is the one closest to the nucleus.

Think of the permitted orbits in hydrogen as being analogous to the steps of a ladder with the lowest-energy orbit being the ground. An electron in a higher-energy orbit, like a rubber ball on one of the steps of the ladder, is only partially stable. If the ladder is bumped, the ball bounces from one step to another down the ladder until it reaches the ground. In like manner, the electron resides in a higher-energy orbit only temporarily before finding its way back to the lowest-energy orbit or the ground state. In the ground state, the electron is stable indefinitely, as is the rubber ball when it is lying on the ground. But if the electron normally exists in the ground state, how does it gain the energy necessary to exist in one its higher-energy orbits?

11.2. PHOTONS IN THE BOHR ATOM

When a hydrogen atom absorbs energy, it is said to be excited, and the single electron in the atom appears in one of the outer orbits, which have successively higher energies than the lowest orbit (ground state). The electron's change (up or down) from one permitted orbit to another is called an electron transition. A hydrogen atom in a gas may acquire the internal energy that excites its electron by:

- random thermal collisions with other gas atoms;
- collisions with subatomic particles such as free electrons; or,
- absorption of a photon traveling through the gas.

As representative of these three processes, let us consider the excitation of hydrogen by photon absorption. Of all the photons encountering an atom, only those possessing an amount of energy equal to the energy difference between a higher-energy orbit and the one in which the electron is located will be absorbed with the photon's energy being used to excite the atom. For example for hydrogen, it takes 10.2 electron volts (eV), or 1.63×10^{-11} erg, of energy to raise an electron from the ground state to the next higher-energy level. Photons with energies below 10.2 eV can not be absorbed, and consequently the electron can not be excited. Photons with energies in excess of 10.2 eV cannot raise the electron to the second energy level, but they may, if they have the right amount of energy, excite the electron to even higher-energy levels.

How long does an excited atom remain that way? For an excited hydrogen atom, in about a hundred-millionth of a second it rids itself of any energy in excess of that of the lowest-energy orbit by emitting the excess energy as one or more photons. Like the rubber ball bouncing down the steps of the ladder, the electron drops in succession into one or maybe several lower-energy orbits on its way to the ground state. In each energy level in which the electron appears, other than the ground state where it can reside indefinitely, the electron remains again about 10^{-8} s. With each downward transition, a photon of electromagnetic radiation is emitted in which the photon contains the energy difference between the two orbits involved in the transition. The greater the energy difference, the greater is the amount of resident energy in the photon, and consequently, the shorter is the photon's wavelength.

In addition to the model of the atom that represents it by its electron orbits, we can make a model using the energy of each allowed electron orbit. The number of energy levels corresponds in a one-to-one fashion with the number of electron orbits. The distance between successive electron orbits increases with higher orbit numbers, but the differences in energy between successive levels grows smaller as the orbit numbers increase.

Suppose a hydrogen atom is excited such that its electron is in the third energy level. The electron remains there about 10^{-8} s and then may de-excite directly to the ground state, emitting one photon. The photon would have a wavelength of 1026 A, which corresponds to the energy difference between these two energy levels. Alternatively, the electron may de-excite to the second energy level, emitting a photon with a wavelength of 6562 A and then de-excite about 10^{-8} s later from the second energy level to the ground state, emitting a photon with a wavelength of 1216 A. The total energy emitted in both cases is the same, but the wavelengths of the photons that result and hence the spectral lines differ.

The hydrogen-line spectrum in the visible region, known as the Balmer series, is prominent in the absorption spectra of most stars. It arises from electron transitions originating on the second energy level of hydrogen. In the same way, all possible transitions from the ground state are known as the Lyman series, which is in the ultraviolet part of the electromagnetic spectrum. Those transitions from the third level up to higher energy levels constitute the Paschen series, which is in the infrared region and so on for the remaining series, whose lines appear in the far infrared on out to the microwave region.

For hydrogen, each series of spectral lines comes to a limit toward shorter wavelengths. The uppermost energy levels, representing the electron's highest energy orbits, crowd together toward a series limit, which represents the point beyond which the proton can no longer bind the electron. Once an electron has an energy beyond the series limit, it leaves an atom, that is, the atom becomes ionized. An ionized hydrogen atom cannot absorb or reradiate energy in the form of discrete lines until it captures another electron. Protons capture electrons because of their electrical force of attraction between themselves being positively charged particles and the negatively charged electron.

In the Bohr atom, besides limits on the size of electron orbits, there is a limit to the number of electrons that may occupy a given orbit. These allowed orbits with a prescribed number of electrons in them are called electron shells.

In general, as one goes through the periodic table, electrons are added to balance the number of protons in the nucleus by filling the shells from the one closest to the nucleus outward. In hydrogen there is one electron in the innermost shell, which has room for a maximum of two electrons. Helium's two electrons fill, or close, the shell, so that for the element lithium the third electron must start a new shell, which is the next innermost. In the second shell there is room for only 8 electrons; in the third, 18 electrons; in the fourth, 32; and so on.

Considering the 92 naturally occurring elements, we find that there are 92 distinct configurations of electron orbits, that is, each element has a unique set of energy levels. Consequently, the wavelength of the spectral lines originating from electron transitions between various energy levels is also unique for each element, a clear fingerprint of the element.

The amount of energy needed to ionize an atom varies from one element to the next depending on the number and "position" of the electrons. For example, to remove the outermost electron from helium takes five times as much energy as it does to do the same for sodium. Also, for a given element, each additional ionization takes more energy to free an electron from an inner orbit than from an outer one because the inner one is more tightly bound to the nucleus. Thus, with carbon, for example, it takes more than twice as much energy to remove the second electron than it does to remove the first electron, 4 times more for the third electron than for the first; almost 6 times more for the fourth electron, 35 times more for the fifth electron, and 44 times more for the innermost sixth electron.

Multiple ionization of carbon brings a corresponding readjustment of the energy levels because of the altered electrical attraction between the positive nucleus and the reduced number of electrons. Altering the energy corresponding to each allowed orbit produces different spectral lines with each succeeding ionization of the carbon atom. Consequently, we see not only a different set of wavelengths in absorption or emission spectra for different unionized elements, but also for the same element after each ionization. That is, the spectrum of singly ionized carbon differs from that of neutral carbon, the spectrum of doubly ionized carbon differs from that of singly ionized carbon, that of triply ionized carbon differs from that of doubly ionized carbon, and so on.

Typical reactions by nonscientists to

Bohr's conceptual scheme were to wonder whether it is about reality, since his theory is quite remote from their everyday experience. For a moment, let us consider Bohr's theory as a process in science. The photon concept by Planck and Einstein was developed earlier on the basis of phenomena other than the emission of discrete wavelengths by atoms. What Bohr did was to combine Einstein's photon concept and that of Rutherford's nuclear atom in a bolder conceptual scheme to show that the internal behavior and structure of atoms can account for the observed emission, or absorption, spectra of atoms. Although additions have been made to Bohr's theory over time to broaden it into a mechanics of the atom called quantum mechanics, the basic concepts are still those of Bohr.

There exists a vast collection of data on emission and absorption spectra of atoms, ions, and molecules. Although Bohr's theory may provide a framework for understanding all of this data, why does one say that Bohr's theory "explains" this data? Let us remind ourselves that explanations are achieved in science when scientists are able to fit new physical phenomena into the body of science using a mixture of existing concepts and a minimum of new ones, all of which are consistent with their preconceptions about how nature operates. If the process of accounting for these new phenomena includes mathematical laws and principles, as does Bohr's theory, then all the better and we have a sound basis for making predictions about new but similar phenomena. In that sense, Bohr's theory is an explanation of nature.

Fine, but is Bohr's theory about reality? Does the "real" world really look like Bohr's miniature planetary system? That we even question its reality, is only because Bohr's theory is based on a model that is unlike any mechanism with which we have had immediate experience. Note that we did not question the reality of Newton's gravitational model of the Solar System, yet with the exception of falling bodies we have no immediate experience with one body orbiting about another body because of something called gravity. Yes, we can build mechanical models, strings, gears, etc., of orbiting bodies, but we can not do laboratory experiments in which gravity is solely responsible for orbital motion. Why then is Newton's theory anymore about reality than is Bohr's theory? The answer is that the business of science is to make the unfamiliar familiar by bringing the unfamiliar into the structure of explained experience. After so doing, the human mind with time and repeated exposure, makes the unfamiliar familiar and accepts the unfamiliar as "real." Newton's theory is familiar to us while Bohr's theory is unfamiliar, and whether or not it is about reality depends on your experiences with it. Given sufficient exposure to Bohr's theory, in time it will in your mind, as in the minds of physical scientists, be seen to capture reality.

11.3. INFORMATION IN CONTINUOUS SPECTRA

At the same time physicists were working on the structure of the atom and how atoms produce discrete emission spectra, work had been underway for sometime on continuous spectra. Continuous spectra were instrumental in the development of the photon concept. From those considerations came fundamental tools for astronomers in their pursuit of understanding of the nature of stars.

All material objects radiate and absorb electromagnetic radiation; the wavelength region and the amount of energy depends generally on the object's temperature and physical state. Varying the temperature in laboratory experiments and from theory, physicists in the nineteenth century analyzed how various bodies emit and absorb radiation. From this work they developed the concept of an idealized radiator called a blackbody.

A blackbody is an imaginary body that, when cool, absorbs all wavelengths of radiant energy falling on its surface so that it is black in color; when hot, the blackbody emits energy with 100 percent efficiency. In reality, matter is generally less than 100 percent efficient when it radiates.

At room temperature, lampblack, a finely powdered black soot, is very close to being a blackbody because it absorbs almost all the radiation incident upon it and reflects very little.

For our purposes, the most important feature of blackbodies is the way in which their emitted radiant energy is spread out in wavelength, or the spectral energy distribution. Physicists have found that the distribution of energy depends only on a blackbody's temperature and not on its chemical composition or any other aspect of it. Note how the amount of radiant energy emitted by a blackbody varies with wavelength in a very recognizable way, even for different temperatures. The emission of radiant energy, or the brightness at each wavelength, covers a continuous range of wavelengths so that the spectrum of a blackbody is a continuous spectrum.

In 1900, the German physicist Max Planck derived a mathematical expression, now called Planck's Law, which describes the distribution of brightness in the spectrum of a blackbody. There are two other distinguishing characteristics of the spectrum of blackbody radiation, these are:

- the energy emitted by the blackbody is greater at every wavelength as the temperature increase, thus the total amount of radiant energy emitted increases with increasing temperature, which is known as the Stefan-Boltzmann law; and,

• the greatest amount of radiation is found toward shorter wavelengths, blue end of the visible spectrum, as the temperature increases, and this is known as Wien's displacement law.

The significance of the blackbody-radiation laws, that is, Planck's Law, the Stefan-Boltzmann Law, and Wien's Law, is that when bodies emit electromagnetic radiation like that of a blackbody, they do so because they are hot. Fortunately, the radiation emitted by stars tends to be much like that emitted by a blackbody. Thus the blackbody-radiation laws are powerful diagnostic tools for measuring the temperatures of stars as thermal sources of radiation. We shall use this fact in our study of the Sun and stars in this and the following chapters. For the study of bodies that emit radiation not because they are hot, called non-thermal sources of radiation, but because of some selective physical processes, the blackbody-radiation laws are of no use. Some everyday examples of thermal sources of radiation are an incandescent light bulb, the burner on an electric stove, and the flame of a cutting torch. Examples of non-thermal sources are a fluorescent light, lightning, and a television screen. However, stars are the examples of thermal sources in which we are most interested.

11.4. RADIATION FROM THE SUN

Let us use the blackbody radiation laws and our knowledge of the absorption of radiation by atoms and ions to consider the outer layers of stars, such as the Sun. For it is these outer layers that are the sources of the radiation we receive from stars.

The Sun has a mass roughly 300,000 times that of the Earth and it emits by Earth standards an immense quantity of electromagnetic energy. Only about half a billionth of the Sun's radiation is actually intercepted by the Earth as it passes out through the Solar System.

How is this radiant energy outflow measured? Starting in 1980 with the launch of the Solar Maximum Mission spacecraft, astronomers could directly measure radiation falling on a unit area just outside the Earth's atmosphere within a certain time, a quantity known historically as the solar constant. The name is now known to be somewhat of a misnomer inasmuch as spacecraft measurements revealed that the solar constant varies by as much as 0.1 to 0.3 percent in the time span of a week or two. However, if we ignore these small variations and their consequences for the moment, we can proceed to find the Sun's rate of emission of radiant energy over all wavelengths, or its luminosity. If we multiply the average solar constant by the surface area of a sphere whose radius is the Earth's mean distance from the Sun, that is 1 AU, we obtain the rate at which solar radiation flows out from the Sun in all directions. This number must also be the rate at which electromagnetic energy is radiated away from the Sun's surface or its luminosity, whose value works out to be about 4×10^{33} erg/s. Approximately 40 percent of this energy is in the visible part of the spectrum, 50 percent in the infrared region, and the remaining 10 percent in the ultraviolet.

This flood of radiant energy, that is, the solar luminosity, comes from what appears to be the surface of the Sun. It is not in reality a distinct surface, like that of the Earth, but a layer of gas several hundred kilometers in thickness called the photosphere. Dotted here and there with sunspots, the photosphere is actually only the lowest visible level of a much more extensive solar atmosphere. Lying above the photosphere is a transparent, tenuous layer called the chromosphere, which is several thousand kilometers thick. This is topped by an even more rarefied layer called the corona, which extends millions of kilometers out from the Sun in all directions. These three regions, which gradually merge into one another, can be distinguished from each other by their different physical characteristics.

Why do we see only the photosphere in visible light and not the chromosphere and corona? The reason is that visible light coming from the chromosphere and corona is usually too weak to be seen against the much brighter photosphere. But they are visible outside the Sun's limb, or edge, during a total eclipse of the Sun, when the Moon covers the photosphere. The gases of the chromosphere and corona are transparent in the visible part of the spectrum, and photons from the photosphere pass directly through these layers. The chromosphere and corona can be observed directly in short-wavelength regions of the electromagnetic spectrum that is, ultraviolet and X-ray, or in the long-wavelength radio region. This is because the photosphere is not very bright in either of these wavelength regions in comparison to the chromosphere and corona. In part the chromosphere and corona are brighter than the photosphere in the short wavelength regions because the chromosphere and the corona are hotter than the pho.

Below the photosphere the solar gases become opaque and hide the Sun's interior from our view. Although the interior of the Sun cannot be studied directly, astronomers have devised mathematical models of the Sun's internal structure by which these hidden regions can be studied. From such studies we find that the Sun's luminosity is the result of hydrogen fusion, which is the conversion of four hydrogen nuclei into one helium nucleus, occurring close to the Sun's center. This energy, when first released in the deep interior, is chiefly in the form of gamma-ray and X-ray photons. As these photons work their way toward the surface, various atoms and ions absorb and reemit them, which tends to shift the wavelengths of photons from short values toward longer ones.

The emerging photons finally reach those layers lying some 100,000 km below the photosphere, a region known as the convection zone. Through this region the movement of energy is like that in a heated room, where cold, heavier air descends to be reheated near the floor and then rises, carrying heat toward the ceiling. In the Sun, hot gases bring thermal energy up from the bottom of the convection zone to its top lying just below the photosphere, and from there cool gases return to the bottom of the convection zone to start the cycle again.

At the surface most of the radiant energy that left the deep interior hundreds of thousands of years earlier is now in the visible part of the spectrum, the ordinary sunlight that we observe here on Earth about 8 minutes after it leaves the Sun.

The spectrum of the visible solar disk is a continuous band of colors from red to violet crossed by many absorption lines. That is, the Sun's photospheric spectrum is actually an absorption spectrum. In 1814, German physicist, Joseph von Fraunhofer (1787-1826), mapped nearly 600 of the most prominent lines. He designated the strongest absorption lines by capital letters, beginning with the letter A in the red and going to the letter K in the violet. Since the photospheric spectrum is an absorption spectrum, we can interpret these lines according to Kirchhoff's Third Law of spectral analysis: The radiation coming up from the interior of the Sun has a continuous spectrum; as the radiation passes through the photosphere, certain wavelengths are absorbed by atoms and ions of different chemical species in the photosphere's cooler layers and in the adjoining low chromosphere, causing the observed dark lines. The uninterrupted-wavelength regions, between absorption lines, are those of continuous radiation that pass into space without being absorbed.

The temperature of the solar photosphere, or the Sun's surface temperature, is an important property of the Sun, or for that matter any star. It is a measure of the rate at which radiation is emitted by a star that is the star's luminosity. To find the Sun's surface temperature, we can utilize the three methodologies implied in the Stefan-Boltzmann Law, Wien's Law, and Planck's Law. As discussed above, these laws characterize the radiation emitted by blackbodies, but because the Sun is not exactly a blackbody, the temperatures derived from these laws differ slightly; they yield an approximate value of 5800 K. We know that stars can not be precisely blackbodies because the spectra of the radiation from their photospheres would have to be continuous, but simple observation shows us that the spectra of stars are absorption spectra.

One means of demonstrating how a value for temperature is derived, applies Planck's Law, in which the amount of energy in the solar spectrum, measured at a number of wavelengths is compared with that emitted by a blackbody. In such a comparison, we can see that the 5800 K blackbody energy curve is a reasonable approximation to the way energy is distributed in the Sun's continuous spectrum.

The important question is whether this 5800 K surface temperature means the photosphere has but a single temperature, that is, it is constant throughout the entire photosphere, or the surface temperature is actually an average over the thickness of the photosphere. To answer that question let us consider what we actually observe in the case of the Sun.

The Sun's limb, or edge, looks sharp-edged to the naked eye because the layers responsible for emission of white light are too thin to be resolved. They are several hundred kilometers thick, whereas the typical resolution size for 1 second of arc seeing corresponds to about 750 km. Either to the naked eye or in large solar telescopes, the Sun's limb looks darker than the center of the disk. Why is this so, since a shiny aluminum sphere does not appear darker near its edge? At the Sun's edge we are viewing a succession of photospheric layers obliquely, seeing light that comes only from the highest layers of the photosphere. Because the higher layers emit less radiation, we infer from the Stefan-Boltzmann law that they must be cooler, as is evident from the blackbody energy curves. Radiation visible to us from the center of the Sun's disk, however, comes from deeper, hotter layers and is more intense. Thus it is obvious that the temperature declines outward through the photosphere. From this fact astronomers can determine the decrease in temperature and density through the photosphere and use these data, for example, to determine the abundance of the chemical elements.

By measuring the wavelengths of absorption lines in the photospheric spectrum, astronomers have identified in the Sun nearly 70 of the 92 naturally occurring elements and about 20 molecules. The identifications are made by comparing wavelengths of lines in the solar spectrum with those obtained from laboratory analyses of spectra of elements. However, the presence of absorption lines of particular elements only confirms that the element is present, but does not provide the element's abundance directly. The few elements for which absorption lines are missing from the photospheric spectra, mostly the heavier ones, are probably also present in the Sun's atmosphere, but they are either not abundant enough to be detected spectroscopically or their spectral lines are not in the visible or ultraviolet regions, the only regions thoroughly explored.

In order to determine the abundance of elements in the photosphere, we must combine our theoretical knowledge of the probability that an atom will absorb radiation at the wavelength in question at a specified temperature with measurements of the line's darkness and width. In the next section, we shall discuss such measurements for other stars besides the Sun.

Rotation is a common characteristic of most objects in the Universe. Like the planets, ignoring Venus and Uranus, the Sun rotates counterclockwise as seen from north of the orbital plane of the planets. It does not rotate as a solid body, as does the Earth and other terrestrial planets, which is not surprising because the Sun is wholly gaseous. For the atmosphere, primarily the photosphere, the period of rotation progressively increases from 25 days at the solar equator to about 36 or 37 days at the poles. This behavior, called differential rotation, is similar to that which we found for the Jovian planets Jupiter and Saturn. The differential rotation of the Sun has taken on renewed importance in the last decade after it was realized that the interaction between rotation and convective currents below the Sun's surface generates the magnetic fields that are responsible for the host of observed surface activity, such as sunspots.

To find how long it takes the visible layers of the Sun to complete one rotation we measure the travel time of sunspots as they are carried across the disk. Another method, applicable to all solar latitudes, as sunspots rarely appear beyond 40 degrees on either side of the solar equator, measures the difference in Doppler shift for spectral lines in the radiation from opposite limbs of the Sun. A Doppler shift is present because the eastern limb of the Sun rotates toward and the western limb away from the Earth. The velocity relative to the Sun's center is found to be about 2 km/s at the equator. Dividing the distance traveled in one rotation, that is, the circumference, by the velocity gives the time for one complete rotation at the equator, approximately 25 days.

The differential rotation of the photosphere probably does not extend much deeper than the bottom of the hydrogen convection zone. However, evidence suggests that the layers under the photosphere may rotate faster than the photospheric layers do. An additional fascinating possibility derived from long-term studies of sunspots and Doppler shifts is that the rotation of the photospheric surface is not uniform in time. The surface appears to speed up and slow down by several percent or a few tens of meters per second over the 11-year sunspot cycle.

Chapter 12
The Outer Layers of the Sun

We asserted very early in this book that one facet which makes scientific discover possible is the possession of an intuitive feeling for nature, particularly for its quantitative aspects. The Sun, being the closest star to us, has fashioned and refined that intuitive feeling that astronomers possess for the nature of stars when they approach their study. Because other stars are immensely far away, astronomers, extrapolating and feeling their way forward by analogy with the Sun, have pursued a study of stellar surface features even when they could not directly observe their surfaces. Hence, the Sun has been a vital bridge to the world of stars and the variety of phenomena that must be occurring in their outer layers. Even today, in the largest telescopes, stars are still basically just points of light. However, over the last few decades, technology has made it possible to study the outer layers of other stars at a level of detail that is still crude compared to what we can achieve for the Sun, but is, nevertheless, a dramatic improvement over earlier achievements. Let us begin this chapter by surveying the Sun's surface layers.

12.1. STRUCTURE IN THE SUN'S PHOTOSPHERE

The solar photosphere is a transition layer from the invisible interior to the external environment surrounding the Sun. Through this 500-km thick layer temperature declines outward by several thousand Kelvins. The average temperature, or surface temperature as it is known, is about 5800 K as derived from the observed luminosity by using the blackbody radiation laws. The reason that the solar photosphere approximates a blackbody is because it quickly becomes opaque as we probe deeper into it. Photospheres of other stars also approximate blackbodies to a greater or lesser extent than the Sun. However, when we look closely at the solar photosphere, we find a far more detailed structure than one might presume from something as bland as the blackbody radiation laws.

Even in photographs of the solar

photosphere taken in white light a number of features are evident. We have already noted that the photosphere darkens toward the limb; near the limb we can also see bright patches called faculae. And in high-resolution white-light photographs, one can see that the entire disk is covered at all times by small, bright features separated by dark lanes called granules.

Remarkably clear pictures of the solar surface in narrow wavelength ranges were made by a telescope mounted in the Skylab station. Fine details on the surface are not blurred by the Earth's atmosphere at the altitude that the station was orbiting. Such high-resolution photographic studies reveal a potpourri of bright granules with dark inter-granular lanes; these give the surface the honeycombed appearance. Time sequences of photographs show granules forming, disappearing, and re-forming in cycles lasting several minutes. At any given time the whole photosphere is broken up into better than 4 million granules, each occupying roughly 1 million km^2 of the surface. Obviously the photosphere is not a uniform layer of gas; the temperature of the photosphere must vary not only in depth but also laterally across the face of the Sun.

The granules are cells of gas with characteristic diameters of 1000 km and lifetimes of several minutes. From the bright center of the granule to the darker intergranular region, the brightness variation corresponds to a temperature difference of about 200 K. Photospheric granules are a form of convection resulting from the upwelling of unstable convective elements from the hydrogen convection zone below the photosphere. As evidence of this convective exchange, spectral lines from the bright centers are Doppler-shifted toward the blue, that means coming toward us, and those from the dark inter-granular regions are shifted toward the red, meaning they are moving away from us. These Doppler shifts indicate that the bright centers are hot rising gas moving at a few tenths of a kilometer per second that radiates its excess energy and then forms the cool sinking gas in the dark inter-granular lanes.

In 1960, vertical oscillatory motions were detected in and above the solar granulation which possess a period of almost exactly 5 minutes with velocities of about 0.5 km/s. Thus the layers above the hydrogen convection zone are moving up and down with respect to the mean position of the photosphere and low chromosphere. The typical excursion is on the order of 50 to 100 km. The motion seems to be organized over a few thousand kilometers and has been reported to cover areas as large as 50,000 km, with roughly two-thirds of the solar surface experiencing oscillations at any given moment. In 1984, the Sun's closest stellar neighbor, Alpha Centauri, was also shown to be also undergoing 5-minute oscillations. It now appears that the 5-minute oscillation is but one extreme in a range of oscillations, with a 160-minute oscillation as the other extreme. Thus the Sun quivers much like a bowl of gelatin. The consequence of these oscillations is an understanding of the interior structure of the Sun.

Dark features on the Sun have been reported for at least 2000 years. Several sightings per century are contained in ancient Chinese records. Although not the first sightings recorded in Europe, Galileo's telescopic observations in 1610 provided the first details on sunspots, the most conspicuous of a number of transient phenomena to be found in the solar atmosphere.

A typical sunspot has a cellular structure with a dark center, the umbra, surrounded by a grayish filamentary region, the penumbra. Although sunspots emit radiation, the umbra looks dark because it is seen against an even brighter photospheric background, whose temperature is some 1800 K higher. The umbra is about one-fourth as bright as the photosphere and the penumbra about three-fourths as bright.

Sunspots develop in a matter of hours as small pores in the inter-granular region of the photosphere. They grow rapidly, and they generally form in clusters, marking a sunspot group, whose orientation is approximately parallel to the solar equator. Each end of the group is often dominated by a large spot surrounded by smaller spots. The very largest groups may cover up to one-fifth the solar diameter. Sometimes a sunspot group persists for several months, but a typical lifetime is about 1 week. A typical large spot in a group is some 10,000 km across; exceptional ones are 50,000 km in diameter, or about four times the diameter of the Earth. In a week or so this large spot builds to its maximum diameter; then its size slowly declines. Individual spots in a sunspot group undergo slow changes from day to day while they maintain their association.

More than two centuries ago it was discovered that sunspots come and go in a roughly 11 year cycle. The sunspot number shows many cycles of this 11 year variation. From any plot of the sunspot number, it is obvious that the heights of successive maxima are unequal, and the interval between successive peaks or troughs is not constant. Thus, the 11-year period is a very rough average. Each sunspot cycle opens with spots forming at latitudes around 35°N and 35°S, and as the cycle progresses, spots form closer to the equator in both hemispheres. The maximum number of spots form when sunspots are forming at latitudes around 25° in both hemispheres. When the last spots of a cycle are forming near 5°N or 5°S, a few spots again form at latitudes around 35ºN or 35ºS herald the beginning of a new cycle.

Immense arching, curved features are observe around sunspot groups. They are structures of gas whose shape is determined by curved magnetic field lines, since sunspots are known to be the centers of intense magnetic fields. Astronomers know this because of the Zeeman Effect. It was first identified in the absorption lines of sunspot spectra by George Ellery Hale (1868-1938) at the Mount Wilson Observatory in 1908. In the Zeeman Effect, the strength of the magnetic field can be determined from the separation of components of absorption lines, while the direction of the field is shown by the sense of polarization of these components.

The leading spots of a spot group, in the forward direction of the Sun's rotation, are opposite in polarity from the following spots. Opposite polarities are like those of the north and south ends of a bar magnet. The unit of measure of the intensity or strength of the magnetic field is called a gauss. The measured field strength in sunspots exceeds that of the Earth's field by several thousand times, being several thousand gauss. In the Sun's northern and southern hemispheres, polarities of leading and following spots are also opposite to each other. That is, the polarity of the leading spot in the northern hemisphere may be north-seeking in one sunspot cycle, whereas that for the southern hemisphere is south-seeking. Then in the next sunspot cycle the leading spot will be south-seeking in the northern and north-seeking in the southern hemisphere. Thus magnetic field polarity reverses in both hemispheres in succeeding sunspot cycles.

As a whole, the Sun does not have a general magnetic field like that of the Earth. But by averaging over small localized and intense magnetic fields, one gets the impression of a general field a few times stronger than the Earth's field, or several gauss. Magnetic measurements have been made for the Sun fairly regularly over the last 40 years. From which it has been found that magnetic fields in the polar regions reverse polarity near the time of sunspot maxima. Thus what appears to be a general field is probably the accumulation of surface fields that have drifted into the polar regions.

12.2. CHROMOSPHERE OF THE SUN

During a total eclipse of the Sun, when the Moon has just covered the photosphere, a thin, about 2000 km pinkish fringe of light, called the chromosphere, appears beyond the Moon's edge. The chromosphere gets its reddish hue from the emission of radiation in the red-colored alpha line of the Balmer series of hydrogen. Projecting from the chromosphere here and there are rosy arches and loops of gas called prominences, which may extend 100,000 km or more into the overlying corona.

For the few seconds of an eclipse when only the chromosphere is exposed, the photosphere's normal absorption spectrum is no longer visible, and we see an emission spectrum called the flash spectrum. It is the spectrum of light originating in the chromosphere. Many of the emission lines match the wavelengths of the absorption lines, but among the exceptions is a bright yellow line produced by helium. Why do helium emission lines appear in the chromospheric flash spectrum but not in the photospheric absorption spectrum? The reason is the chromosphere's higher temperature, up to 30,000 K at the highest level, and lower density. Neutral helium can be excited to emit radiation only when the temperature is greater than 10,000 K. And the appearance of ionized helium lines requires temperatures in excess of 20,000 K. From this we conclude that the temperature must rise rapidly from the top of the photosphere up through the chromosphere.

Chromospheric events can be monitored, even when there is no eclipse, by photographing the chromosphere in monochromatic light. A single-wavelength photographic device, the spectroheliograph, can make pictures of the solar disk in the residual light of an absorption line, such as the red hydrogen alpha line or the violet K line of singly ionized calcium. Such a picture is called a spectroheliogram.

The reason for choosing the residual light of a strong absorption line is that most of the photons from the photosphere have been absorbed and the remaining ones are those coming from the low chromosphere. Photons in the continuous spectrum come from the bottom of the photosphere, whereas photons in stronger and stronger absorption lines come from higher and higher up in the photosphere and the low chromosphere. By choosing absorption lines of different strengths, we can photograph different levels of the Sun's atmosphere.

Photographs of the chromosphere show us a very different view of the Sun from what we see in the photosphere. Bright patches in the chromosphere, called plages, are seen to overlie photospheric sunspot groups. Hotter and probably more dense than the normal chromosphere, with a spatially averaged magnetic field of a few hundred gauss, plages are typically 10 times larger than the sunspot group lying below them. Plages are nearly always found above regions in the photosphere in which a strong magnetic field exists, and they always appear before the spots form. Their usual life span is about 40 to 50 days, during which several spot groups or none at all may form. The plage areas do not look the same in the red hydrogen alpha line and in the violet calcium K line, but they obviously mark the same general region of the chromosphere.

In a violet calcium K line spectroheliogram one sees a network of bright gas surrounding dark cells in addition to the larger plages. This chromospheric network constitutes the boundaries of large-scale convective cells, known as supergranules, which are seen as Doppler shifts in photospheric pictures. Supergranules derive their name because of their resemblance to convective motions and the fact that they are typically an order of magnitude larger than granules; beyond their names, granules and supergranules have little in common. Apparently the chromospheric network is the locus of very intense and highly localized magnetic fields that are concentrated on the boundaries of the supergranule cells by the motions of gas.

In short-exposure spectroheliograms, the chromosphere appears stippled with a myriad of jet-like spikes of gas, called spicules. Spicules rise rapidly from the chromospheric network, attaining typical heights of 10,000 km, and then they fade away or collapse in several minutes. At any instant, 250,000 of them may cover a few percent of the Sun's surface. They too outline the boundaries of the supergranule cells. Red hydrogen alpha line spectroheliograms contain short dark features, like blades of grass, which are thought to be spicules outlining the bright interiors of supergranule cells. Skylab photographs suggested that the chromosphere may have granule-like features much larger than those observed in the photosphere.

Solar flares are perhaps the most complex of the Sun's transient phenomena. They vary in size, brightness, and behavior, and flaring activity is most common when sunspots are most numerous. A solar flare may suddenly erupt as an intensely bright area in a chromospheric plage that will send material hurdling outward into space. Emitting radiation strongly throughout the electromagnetic spectrum, flares rise to great brilliance in several minutes and then fade in half an hour to several hours.

Solar flares result from the sudden release of energy stored in coronal magnetic fields above sunspot groups. That is, the radiant and thermal energy, which for a large flare may amount to as much as 10^{32} erg, comes from energy stored in twisted magnetic field lines. If we think of magnetic field lines as being like the rubber band of a toy airplane, then the storage of energy is analogous to winding the rubber band. There seems to be a limit to how much twisting field lines can tolerate. When that limit is reached, it is only a matter of seconds to minutes before the energy stored in the field is released as thermal energy of motion, just as releasing the rubber band turns the propeller of the toy airplane.

In flares, free electrons are accelerated up to velocities of about half the speed of light. As these energetic electrons collide with ambient gas, they share their kinetic energy and heat the gas to a few thousand degrees in the chromosphere and to as much as 20 million K in the low corona. This heating phase may last from seconds to minutes and is responsible for the X-ray, ultraviolet, and visible radiation emitted by flares. Some high-energy electrons pass out through the corona, where they excite successive layers to emit radio-frequency radiation.

A major flare is the most energetic of all solar events, equaling in power 100 million hydrogen bombs with a yield of a 100 megatons each. The ultraviolet and X-ray radiation arriving at the Earth can disturb the ionosphere, persisting sometimes for hours. Over the few days following the flare outburst, subatomic particles may spiral into the Earth's polar regions, causing brilliant auroral displays and radio blackouts. All these events, which do not necessarily occur with every flare, partially distort the Earth's magnetosphere, generate geomagnetic storms, and induce severe electrical power surges. Solar flares are one of the few astronomical events that directly disturb our terrestrial environment.

12.3. CORONA OF THE SUN

The corona is that region of the solar atmosphere lying above the chromosphere. During a total eclipse, it is the large halo of white, glowing gas extending out a few solar radii, that is millions of kilometers, beyond the dark limb of the Moon. At times when an eclipse is not in progress, specially designed refracting telescopes, called coronagraphs, that block out light from the photosphere are used to observe the corona.

Compared with the many hours of almost continuous surveillance accumulated in satellite studies over the last 20 years, eclipse studies have yielded not more than a few hours of observations because an eclipse seldom lasts longer than a few minutes. Even coronagraphic studies, which have greatly increased observing time, cannot match the time coverage and resolution of satellite observations from above the atmosphere. In February of 1980, the Solar Maximum Mission spacecraft provided an opportunity to keep watch on the Sun during its period of maximum surface activity. However, in November of that year, Solar Max lost its attitude-control system, so that the craft could no longer point its eight instruments at interesting portions of the solar surface. In April of 1984, the crew of the Space Shuttle Challenger was able to wrestle Solar Max into the cargo bay of Challenger for repair. This first successful repair of a satellite in orbit opened a new era for space observatories.

Approximately 30 emission lines have been identified in the visible part of the coronal spectrum, and many hundreds of emission lines are known in the ultraviolet and X-ray spectrum. They originate in highly excited ions of familiar elements, such as iron, from which several to as many as 15 electrons have been stripped in the corona's extremely hot, tenuous gases. It takes temperatures from many hundreds of thousands up to several million degrees to sustain such a degree of ionization.

From millimeter to meter wavelengths there is a wide spectral window in the Earth's atmosphere that lets in radio radiation. The Sun, when quiet and undisturbed, normally emits thermal, or blackbody radiation, which is characteristic of a million-degree corona. When the Sun is disturbed, as when solar flares are occurring, non-thermal radio emission is added to the thermal component, and it can be quite intense.

There are several lines of evidence, besides the coronal spectrum, that confirms a rise in temperature through the chromosphere into the corona. Since heat flows from high to low temperature regions, then clearly energy must be pumped by some mechanism from the low-temperature photosphere to the high-temperature corona. For a number of years astronomers thought that the corona's high temperature resulted from energy carried into the corona by mechanical waves starting in the turbulent hydrogen convection zone below the photosphere.

As evidence grew that the magnetic fields of the photosphere and chromosphere were highly localized and very intense, it seemed hard to ignore the possibility that these magnetic fields extending up into the corona were part of the coronal heating process. When X-ray pictures showed that the corona was divided into active regions and hole regions primarily because of the structures of their magnetic fields, it became readily apparent that most, if not all, of the heating involves magnetic fields. The heating is produced by the direct dissipation of the energy stored in magnetic fields into thermal energy in the coronal gas. The lower parts of the chromosphere, however, are still thought to be heated by mechanical waves.

Eclipse pictures of the corona provide evidence for the importance of magnetic fields in structuring the corona. In white-light photographs one can see that the corona is irregular and structured. Beautiful, long streamers extend outward in the Sun's equatorial regions. Near sunspot maximum the corona is nearly circular, with streamers radiating out in all directions. Near sunspot minimum, the corona extends farther out in the equatorial region and terminates rather abruptly, with short, thin plumes curving out of polar areas.

Because coronal gases are almost transparent, we often are looking through several structures at once in eclipse pictures, which blur the details. This is why direct photographs in X-ray and extreme ultraviolet wavelengths, where we look down on top of coronal structures, are so valuable to the study of the corona. To photograph the corona directly, we must observe in the 10 to 900 A which is in the X-ray to extreme ultraviolet wavelength region, where radiation from the much hotter corona overwhelms the short-wavelength radiation of the photosphere. But X-ray or ultraviolet pictures must be taken from space because the Earth's atmosphere absorbs these short wavelengths. In X-ray and ultraviolet pictures the corona appears highly inhomogeneous and generally asymmetrical, and it varies over time on both short and long time scales.

There are three different types of structural regions that collectively characterize the entire solar corona: Coronal holes and coronal active regions are two of them, and they are reasonably well defined in terms of observational characteristics; the third is the coronal quiet regions, which are not well defined. Coronal holes are regions of slightly lower temperatures and significantly lower densities, with magnetic fields of about 10 gauss whose lines open out into interplanetary space. Rays and polar plumes extend out of them away from the Sun. Coronal holes are also thought to be the source of most of the subatomic particles in the solar wind. Coronal active regions, however, are extremely different from coronal holes. They consist of loop structures. They are somewhat hotter and much denser regions of the corona, and their magnetic fields of about 100 gauss having field lines that loop back into the Sun instead of extending outward as in the case of coronal holes. Between these two extremes are the ill-defined coronal quiet regions, which appear to be something in between. There are also many small, bright points visible in X-ray pictures.

With time-lapse photography of the corona one sees spectacular motions of towering masses of luminous gas, called prominences. Projected against the solar disk, they are the dark, threadlike filaments. Their forms vary from almost stationary, quiescent arches and graceful loops to rapidly moving surges.

The typical prominence is hundreds of thousands of kilometers long and extends several tens of thousands of kilometers above the photosphere. It consists of gas cooler and denser than that in the corona around it. In the more active prominences, gas may rise at rates of hundreds to thousands of kilometers per second, which is sufficient to escape from the Sun.

Frequently; however, matter appears to rain down from the corona in great luminous masses. Apparently, magnetic fields hold up these huge walls of gas against the Sun's gravitational pull. We can see matter flowing along the body of a prominence, following the curving and looping magnetic lines of force.

The mean lifetime of large quiescent prominences is about two to three rotations of the Sun. During sunspot maximum, twenty filaments may appear on the disk; during sunspot minimum, there are typically about four. Prominences always appear to be associated with a plage or sunspot group. In fact, the large quiescent prominences tend to form along the division between regions of different magnetic polarity in plages. The polarity of each side of the plage region is also the same as that of the sunspots in the photosphere underneath it.

12.4. ACTIVITY CYCLE OF THE SUN

During a sunspot cycle, the general level of all activity in the solar atmosphere follows the number of sunspots. Thus a sunspot group seen in white-light photospheric pictures is just the most visible indicator of a large disturbed region in the solar atmosphere called an active region. Such regions can be up to several hundred thousand kilometers in extent. The common bond among these visible features is magnetic fields. The field appears first, followed by a facula in the photosphere and a plage in the chromosphere. This in turn can be followed by a sunspot group, flare activity, and prominences. The precise behavior is somewhat different for each active region, but there is little doubt that the phenomena in different layers of the solar atmosphere are related. The solar cycle of activity, a succession of active regions over a 22-year period, is fundamentally a magnetic cycle. It is a many-year variation in the quantity of magnetic field that emerges in the solar atmosphere.

The Sun's transient activity produces a variety of transient effects here on Earth, which follows a rough cycle in unison with the Sun's cycle. An example of this relation is the increase in auroral activity in the Earth's atmosphere during sunspot maxima and its decrease during minima. Changes in solar activity also affect Earth's weather and climate, but how and over what time scales these effects occur we do not clearly understand. In recent years some interesting discoveries have been made about the constancy of the Sun's activity and its relation to the Earth.

Approximately 1.4 million erg of radiant energy fall on each square centimeter of the Earth every second. The theory of stellar evolution suggests that the Sun's luminosity must have increased by perhaps 30 percent since it began its existence some 4.6 billion years ago. However, as recently as 1975, paleoclimatic evidence concerning long-term temperature variations on Earth showed that any change has been less than 3 percent in the last 1 million years. One can argue that if the solar luminosity had been as much as 25 percent less than the present value, the oceans would have frozen, preventing biological evolution. This suggests that if the solar luminosity has increased significantly since Earth's formation, something about the Earth has compensated for such a change.

Solar Max measurements provide evidence that sunspots and faculae, granulation, and solar oscillations are responsible for as much as 0.4 percent changes in the solar constant over a time scale of a week. And since 1980, the solar constant has steadily decreased by 0.02 percent per year. What longer-term changes exist in the solar constant, if any, are important to our knowledge of the Sun and the Earth.

Another fundamental question is the degree to which the 22-year cycle of solar transient activity repeats itself over long stretches of time. In 1893, E. Walter Maunder (1851-1928), a British astronomer, found in European historical records that very few sunspots were seen in the period from 1645 to 1715, now known as the Maunder Minimum. Within the last several years, Maunder's work has been confirmed and extended by astronomers worldwide. In addition to the absence of sunspots during the Maunder Minimum, very few aurora were observed in Europe, and during eclipses, the corona was also absent or very weak. Virtually no sunspots were reported in

Asia during the Maunder Minimum, even though naked-eye sunspot-sighting reports exist there from as early as 28 B.C. Finally, measurements of the amount of carbon-14, a radioactive isotope of carbon, in tree rings show that during the Maunder Minimum there was an excess of this isotope in the Earth's atmosphere. We believe that the high-energy subatomic particles called cosmic rays, moving randomly through the Galaxy, collide with the nucleus of nitrogen atoms in our atmosphere, converting it to carbon-14. When the Sun is very active, the interplanetary magnetic field is strong, and Galactic cosmic rays are deflected away from the Earth. Thus high levels of carbon-14 in the Earth's atmosphere correspond to low levels of solar activity.

Historical research has shown a correlation among carbon-14 abundance in the tree rings, winter severity, galactic cosmic-ray activity, and solar activity. Periods of colder climate appear to coincide with low levels of solar activity. Evidence exists that at least a dozen similar periods of minimal solar activity, lasting from 50 to 200 years, have occurred over the last 8000 years. Answers as to why the Sun should experience decreases in transient activity are to be found in changes in the magnetic activity which is probably caused by changes in the pattern of convection underneath the photosphere. Consequently the interior of the Sun may not be as constant in its behavior as once thought.

Chapter 13
The Quest for Life

During the course of the history of the Universe, life became possible only after galaxies had formed and their stars had existed long enough to synthesize the heavy atoms found in organisms. The six most important elements for life on Earth, and their percentages by numbers of atoms in human beings are as follows: hydrogen, 61 percent; oxygen, 26 percent; carbon, 11 percent; nitrogen, 2 percent; phosphorus, less than 1 percent; and sulfur, also less than 1 percent. Had the primordial condensate that immediately followed the big bang not cooled rapidly in the first few minutes, life could not have occurred in the Universe because hydrogen would mostly have been converted into helium, leaving little available for stellar nuclear synthesis, and heavy elements would not have been subsequently formed in stellar interiors. Thus an important point is whether enough time has elapsed for stellar evolution to build a significant abundance of the biologically significant atoms from which chemical evolution will flow if given the opportunity. Clearly it has at our position in the Galaxy, or we could not exist.

Biologists are not unanimous on all the factors in the definition of life, but most agree that life is different from nonlife: It evolves or changes, by chance mutations or otherwise, as time goes on while interacting with its environment in a unique way. But most important, life is not a "thing", it is a process made up of an unimaginably large number of complex chemical reactions that collectively produce all the characteristics we associate with life.

13.1. LIFE IN THE SOLAR SYSTEM

The place to begin our discussion is with ourselves, since we are the only system of life for which we have any information. From the Earth we can then move to consider the Solar System; and in the next section, we shall consider the question of life beyond the Solar System.

The thermal habitable zone in our Solar System, that is, the region where life might most likely flourish on a Terrestrial planet because of the presence of liquid water (between 273K and 373K), lies between Venus and Mars. The inner limit is roughly the point where water would boil, and the outer limit is the point where water would freeze. Had Earth formed somewhat closer to the Sun than it did, the greenhouse effect might have become dominant, as it is on Venus, resulting in a hot, sterile surface. And had the Earth formed farther away from the Sun than it did, water would have remained frozen as subsurface ice or polar caps, and the surface of the Earth would somewhat resemble the cold Martian landscape.

Our Earth, therefore, is at the right distance from the Sun for development of an active biosphere. The chemistry of Earth's life is built on the chemistry of the element carbon and the solvent water, supplemented by the biologically important atoms hydrogen, nitrogen, oxygen, phosphorus, and sulfur. Carbon atoms bond easily with other carbon atoms, producing long chains to which other biologically significant atoms can bond. The resulting molecules, whether or not they are part of or a product of living matter, are called organic molecules.

Water, because it can flow readily and remain in liquid form through a range of conditions, is the ideal solvent for many organic compounds. From water come hydrogen bonds, which give structural stability to strings of proteins, nucleic acids, and other long-chain carbon compounds. A liquid water environment and moderate temperature make it possible for such long-chain carbon molecules to form, and they have become the biochemical basis of life as we know it.

Even with this biochemical basis, life would not have been able to develop and sustain itself without proper temperature, a supply of nutrients, self-regulating mechanisms, and the Sun's energy. The Sun is the prime source of energy for driving the chemical reactions in the cycle of life.

How did life begin, and how did it derive from nonlife? Our contemporary ideas on the evolution of life began in 1924, when the Russian biochemist Alexander Oparin (1894-1980) introduced chemical evolution as a necessary forerunner to biological evolution. In 1928, the English biologist J.B.S. Haldane (1892-1964) independently suggested an outline for chemical evolution, which is still the basis of our current understanding.

Earth's cooling and solidifying crust was racked by volcanic activity that presumably vented carbon dioxide, nitrogen, water vapor, hydrogen compounds, and smaller amounts of other molecules that are easily vaporized. These probably formed Earth's early atmosphere. Today active volcanoes discharge large quantities of carbon dioxide, nitrogen, water vapor, some sulfur, and traces of other gases.

Subsequent cooling of Earth condensed water vapor, forming the warm seas and the shallow lagoons and pools that were destined to provide a haven for the development of organic compounds. From this so-called "primordial soup," over a long time, the more complex organic molecules or biological macromolecules, such as proteins and nucleic acids evolved. These were formed by energy from solar ultraviolet and visible light, electric discharges, and heat from radioactivity, volcanoes, and meteoric impacts.

Cells are the basic units of life from which complex organisms, such as human beings, are built. Modern biology has shown that the cell is in general the minimum organized unit of matter that displays the properties of life. Collections of cells make up an organism. The cell has in it a water-based substance surrounding a nucleus, which holds coiled, threadlike strands known as chromosomes. The chromosomes, which transfer hereditary characteristics to each generation of new cells, occur in pairs, with a fixed number in every cell of every species. Human cells have 23 pairs. The nucleus, with its many chromosomes, controls the cell's activities. It contains instructions for manufacturing the specialized cells, such as muscle, bone, or liver cells, and maintaining their functions to keep an organism alive.

Hereditary information is contained in the deoxyribonucleic acid (DNA) molecule, a large complex molecule found in the chromosomes of the nucleus of every living cell. One human cell has about 800,000 DNA molecules. Since there are many varieties of living organisms with many chromosomes, there are many different forms of DNA. DNA has two primary functions. It carries the heredity instructions for manufacturing proteins in the cell, and it passes genetic information on to daughter cells during cell division by making copies of itself, which is called replication.

Essentially, the atoms in DNA are linked like a twisted ladder, or a double helix. The spiral is a right-handed spiral in which each tread is of the same size and at the same distance from the next and turns at a rate of about $30°$ between successive treads. The genetic code, which specifies the hereditary message, is carried on the treads of the ladder and is contained in the sequence of the four different units. Thus the language of heredity is written in an alphabet of only four letters. For each protein potentially capable of being formed, a specific segment of the DNA molecule carries the information by which the 20 kinds of amino acid subunits in the protein are properly ordered during its synthesis.

During cell division, replication of the DNA molecule begins with separation of the two spirals. Each strand of the double helix directs the information of a complementary strand to pair with it. They migrate to opposite ends of the cell before the cell divides. Thus in the daughter cells are two daughter DNA molecules identical to those of the parent DNA molecule in the parent cell. DNA molecules specifically determine the type of organism, for example, human being or elephant.

Biological evolution is accomplished by mutations, which introduce new factors into the genetic message of DNA, by recombination, which is the rearrangement or new association of message units by sex-like processes, inheritance from different parents, and by selection, which is the weeding out of inferior traits in a population through successive generations, not in individuals in their lifetimes. DNA molecules are remarkably stable. Mutations occur about once per gene for every 100,000 cell divisions in most mammals. Evolution in plants and animals is controlled by two major forces which are limits that the environment sets for the organisms and changes in their hereditary material. Organisms are constantly modified by chance mutations, sexual selection, and natural selection. Together these produce members of a species that can survive in a changing environment.

From Earth's fossil record two important points are evident. First, species do not repeat, although essential parts of organisms, such as the eye, may have several independent entries into biological populations. And second, some species have remained essentially unchanged since their appearance, whereas others have shown significant change. Finally, an important consideration is whether or not there is a definite direction to biological evolution.

If we assume that chemical evolution is followed by biological life that evolves toward greater complexity, what can we say about the development of intelligence? The great leap forward for human beings came only in the last few million years, when our humanoid ancestors learned to walk upright, freeing hands for the manipulation of tools. As their brains evolved and their mental capacities expanded under pressure for survival, a collective culture and civilization set human beings apart from other living creatures.

Yet human beings, although the most advanced, are not the only intelligent creatures on Earth. One of the most important characteristics of intelligence is the ability to collect and transfer information. There is a spectrum of this ability among Earth's creatures signifying that they possess varying intellectual capacities. Scientists have found a "language" capacity in chimpanzees, gorillas, and dolphins.

It is unlikely that another intelligent creature would closely resemble us. As we have pointed out, the fossil record does not indicate that evolution repeats itself. Loren Eiseley (1907-1977) noted in a more general sense that we might expect along such lines for the Universe at large: "Life, even cellular life, may exist out yonder in the dark. But, high or low in nature, it will not wear the shape of man. That shape is the evolutionary product of strange, long wandering through the attics of the forest roof, and so great are the chances of failure, that nothing precisely and identically human is likely ever to come that way again."

Among meteorites that have been recovered is a small subgroup of stony meteorites, called carbonaceous chondrites, that contains up to about 5 percent organic molecules. The fact that chondritic meteorites carry organic compounds has been known for more than a century, but only when techniques had been developed for studying lunar rocks could the organic compounds be definitely ascribed to an extraterrestrial origin. Amino acids were discovered in a fresh chondritic specimen that fell near Murchison, Australia, in September of 1969. Extraterrestrial amino acids have since been found in other meteorites, but only a few of these acids are in living cells of Earth organisms. Amino acids from meteorites are almost an equal mixture of right- and left-handed molecules. Amino acids of biological origin are exclusively left-handed, supporting the idea that the meteoritic amino acids are of extraterrestrial origin. If extraterrestrial life produced amino acids, we believe that they would be exclusively either left- or right-handed, so the mixture of both types in the meteorite specimens suggests a chemical, not a biological, origin.

The Moon was the first body outside the Earth on which searches were made for the presence or remains of living organisms. Organic chemists had improved old techniques and developed new ones in anticipation of new findings on biochemical evolution and the origin of life that might come from the lunar samples from the Apollo program. But, alas, nowhere in any of the samples was a significant amount of carbon found. No amino acids, proteins, or nucleic acids were found. Also, no water molecules, either free or chemically bound, were found in these surface materials.

Neither Mercury nor Venus is a very promising site for the development of life. This leaves Mars as the only realistic place for biological exploration. Project Viking, a billion dollar undertaking, was designed in part to search for life on Mars. Two cameras periodically scanned the immediate surroundings to look for any large-scale biological life. But the main biological hunt on Mars was for microorganisms in the Martian soil. In the various Viking biological experiments, all found an unusually active Martian soil, but no evidence for microorganisms. Possibly living organisms developed and evolved in an environment on Mars so different from that on Earth that we did not formulate the experiments correctly, or we are still not interpreting the results properly. At the present time it seems unlikely that Mars does have or has had living organisms on its surface.

Although evidence exists that suggests extensive chemical evolution has occurred either prior to or early in the Solar System's existence, whether or not it is heading toward biological evolution anywhere else besides Earth is not evident. At this point it is much too early to make anything like a final pronouncement about life in the Solar System.

13.2. LIFE IN THE UNIVERSE

The first question we may wish to ask is about the environment, or platform, on which the slow processes of chemical and biological evolution can take place. An initial thought would be to assume that the life process might begin in a planetary system about a star and on a solid surface of one or more of its planets, as did our own development.

If we decided to select stars in the solar neighborhood near which life might be found, how should we proceed? Some criteria have been proposed that could improve the probability of finding stars with planetary systems on which life might exist. The criteria do not argue against the existence of planets in those situations that are eliminated, but rather they try to identify what are the highest probability situations in which planets can support life.

A first criterion for assigning a low probability to a star's having a viable planetary system is the length of time a star is a main-sequence star, which is the longest phase in a star's life. Thus, the hottest stars would be rejected for life-bearing planetary systems because their time on the main sequence is much too short, less than 1 billion years, to permit chemical and biological evolution, which would probably take several billion years as is the case on Earth. Even if the time scale for biological evolution was assumed to be 1 billion years or less, the quantity of high-energy photons, such as X-rays and ultraviolet, emitted by hot stars would probably prevent the formation of complex organic molecules. Although cool stars on the lower end of the stellar temperature scale have a lengthy life span on the main sequence, they are much too cool to support the development of life except in orbits that would be very near the star and possibly unstable, or at least synchronous orbits. For this reason, they are not assigned a very high probability.

Second, we should probably assign a very low probability to the vast majority of binary and multiple stars, about half the stellar population, because the orbits of planets around them might not be stable enough to maintain a planet in a thermally habitable zone. A third criterion would be how rapidly the host star changes, so that all stars in the post-main-sequence stages of evolution are probably not good candidates. If planetary systems had developed and sustained life, then the expansion of stars during red-giant evolution would destroy the thermal environment in which they had existed. And subsequent phases in a host star's existence in many cases may be too short for life to develop a second time somewhere else in that system.

Such arguments leave us with main-sequence stars with surface temperatures from 3700 to 7500 K as the best possibilities for supporting life. If there are approximately 400 billion stars in our Galaxy and 85 percent are main-sequence stars, then there are about 340 billion main-sequence stars. Astronomers estimate that about 90 percent of main-sequence stars have a surface temperature from 3000 to 7500K, or what amounts to about 300 billion stars. One-third of these, or 100 billion, are apparently from 3700 to 7500 K, and one-half of these, or 50 billion, are probably not members of a binary system. Thus there are a large potential number of life-supporting planetary systems. In fact, estimates from the Kepler mission, specifically designed to detect extrasolar planets around distant stars in a specific region of the galaxy, place this estimate as being closer to 100 billion.

To define a habitable zone for life, why do we seek a temperature range in which water is in liquid form? Why not some other organic solvent, such as alcohol or ammonia? First, water is a simple molecule, consisting of just three atoms, of which two, hydrogen, are the most abundant element in the Universe, and the third atom, oxygen, is among the most abundant elements after hydrogen and helium. Second, liquid water can store a great deal of thermal energy before it vaporizes. Thus it acts as a buffer to day-night temperature changes that occur when a planet rotates. Finally, water has a high surface tension, which can help concentrate solids at its boundaries. In a similar vein, carbon chemistry is expected to be a more widespread basis for life than is silicon chemistry or germanium chemistry, both silicon and germanium behave chemically somewhat like carbon. This is so because carbon is far more abundant cosmically than either silicon or germanium.

Another is whether chemical evolution up to the macromolecule stage will be followed by biological evolution, given a suitable environment and sufficient time. Although biological evolution leading to simple life forms is obviously less probable than chemical evolution is, it seems reasonable that biological evolution has occurred many times just from the huge number of opportunities.

Of the 31 stellar systems within 15 light years of the Sun only 3 seem to meet the criteria needed for what some call an ecoshell. These are the main-sequence stars Epsilon Eridani, Epsilon Indi, and Tau Ceti. If we take the solar neighborhood as a representative sample, in which 3 stars out of the 31 stellar systems within 15 light years of the Sun have potentially habitable planets, the average distance between biologically suitable stars is about 17 light years. Therefore, within a radius of 1000 light years, then, we should expect to find somewhat less than 1 million stars having suitable planets harboring some kind of life. Even if only 1 in 1000 of these planetary systems has an intelligent species, that still leaves 1000 sites of intelligent life within 1000 light years. If we conservatively extend this argument, estimating that only 1 million civilizations with a technology at least equal to ours are distributed throughout the Galaxy, the average separation between them would require 600 years either to send or to receive a message, hardly a hurried conversation.

In order to estimate the number of extraterrestrial civilizations in our Galaxy now, we need a definition of the term. We shall define an extraterrestrial civilization as a group of lifeforms technologically capable of and inclined by curiosity to communication with other Galactic civilizations. We can start by making an inquiry into the possible number of communicative civilizations now existing in our Galaxy. An equation developed by Frank Drake (b. 1930), known as the Drake equation, is a general formula expressing the number of such Galactic communities in terms of several factors:

Number of communicative societies = (astronomical factors) x (biological factors) x (sociological factors)

$$N = R_s \, f_p \, n_p \, f_l \, f_i \, f_c \, L.$$

The first of the astronomical factors, R_s, is the number of stars in our Galaxy divided by the life span of the Galaxy. This factor amounts to about 400 billion stars divided by 10 billion years, or about 40 stars per year. It is a rough measure of the rate at which stars form in the Galaxy.

The second astronomical factor, f_p, is the fraction of stars that live long enough for life to develop and to have a planetary system in which it can develop. From our earlier arguments, about 100 billion main sequence stars live long enough for life to develop, but we assume only half of to have planetary systems, or about 50 billion stars. Therefore, f_p is about 0.125, or 50 billion divided by 400 billion.

The third astronomical factor, n_p, is the number of planets in each planetary system suitable for life; it is the product of the average number of planets per planetary system and the fraction that are suitable for life. For the Solar System, that product equals 1. Therefore, let us somewhat arbitrarily choose one planet for each system.

The first of the biological factors, f_l, is the fraction of planetary systems in which life actually appears. The factor $f_l = 0.5$ is arrived at on the assumption that under proper conditions, sooner or later life will take hold, flourish, and evolve into a myriad of thriving forms in every other system.

The second of the biological factors, f_i, is the fraction of evolving systems that evolve at least one intelligent species. We guess that the probability that nature, with say 4 billion years of effort, will create at least one intelligent species on a planet is 50 percent. Therefore, we set the factor f_i equal to 0.5.

Now for the sociological factors. The factor f_c is the fraction of Galactic societies technologically able and willing to take part in interstellar communications; this factor we also guess to be 0.5, that is, a 50 percent chance that the species will develop technological capability and will want to try communicating with other Galactic civilizations.

The factor L is the length of time the civilization continues in its communicative phase. Our own interest in interstellar communication dates back only a few decades in a period of more than 6000 years of civilization.

For the number of intelligent communicative societies, then recognizing a great deal of uncertainty exists in each term,

we obtain
$$N = (40)(0.125)(1)(0.5)(0.5)(0.5)(L)$$
or
$$N = 0.625L.$$

Thus, N is approximately equal to L itself. In other words, the number of communicative civilizations in our Galaxy approximates the average number of years spent in the communicative phase, and the factor L is probably the most uncertain of all to evaluate. Although one may wish to question the assumptions leading to the particular values quoted above, most astronomers believe that they are reasonable. We need to remember also that this result applies only to our Galaxy and does not include the billions upon billions of other galaxies in the Universe.

When we think about the possibilities of our own destruction by nuclear holocaust, by biological disasters from new mutant strains, by changes in the planet's ecology and climatology due to human stupidity and blunders, by terrestrial and extraterrestrial catastrophes, and by other calamities that could befall a civilized society, it is tempting to predict that the moment of civilized glory may indeed be brief in the span of an intelligent species.

Assuming that we have a fair grasp of the values for the product R_s f_p n_p, we can calculate the average separation between communicative societies for various values of the product f_l f_i f_c and L. So what about values that cover a range of reasonable values for the product f_l f_i f_c and L? Note that if a species survives in a communicative phase for only 1000 years, then the length of time for messages to travel the distance between civilizations exceeds the lifetime of the communicative phase. We need to look at only the combinations that present any reasonable chance for an exchange of messages.

Radio techniques had improved so spectacularly after World War II that a few astronomers and physicists privately considered the feasibility of detecting extraterrestrial signals from intelligent life. The subject finally surfaced in the British scientific journal Nature in September of 1959, when the physicists Giuseppe Cocconi (1914-2008) and Philip Morrison (1915-2005) presented logical reasons why efforts should be made to search for interstellar signals generated by intelligent life.

For now it seems more practical for us to listen for signals than to transmit them. Perhaps messages that older, more advanced extraterrestrial civilizations, have been transmitting for centuries have by now reached the Solar System. The most advanced celestial communities could avail themselves of energy sources far more sophisticated and powerful than any we can realize today, perhaps even using the energy output of their parent stars by modulating their light as signals. We may be no more aware of such electromagnetic messages than New Guinea aborigines, who use drums for communication, are aware of the international radio traffic constantly passing overhead.

The first modern attempt in the United States to detect artificial signals from space was conducted by Frank Drake at the National Radio Astronomy Observatory at Green Bank, West Virginia. This undertaking was called Project Ozma, after the legendary princess of the imaginary land of Oz. A 26-m radio telescope was aimed at

Tau Ceti and Epsilon Eridani for 150 hours of observation from May through July of 1960. Although, the effort was not rewarded by finding signals from intelligent beings, the lack of success seems to have nowise diminished interest. For since then, other attempts to detect signals from intelligent beings have been carried out. None has succeeded, but the several hundred stars examined are a very tiny sample of the possible sources. The problem in locating signals is to pick not only the right star but also the right frequency and the right time to observe. Even if a communication were received from another world, it would take great amounts of time to exchange messages, so perhaps the first step in acknowledging contact with an extraterrestrial society would be to transmit a duplicate of the received message back to its source to inform the sending society that its inquiry had been received and recognized as originating from an intelligent source.

The groundwork has been laid for developing the search strategy that could ultimately bring us into communication with extraterrestrial civilizations. This program is known as Search for Extraterrestrial Intelligence (SETI). Working on the premise that a large and expensive radio receiving system is not needed to begin with, SETI would equip existing radio telescopes with low-cost state-of-the-art receiving, data-handling, and data-processing equipment. With this apparatus it should be possible to explore the vicinity of the Sun out to several hundred light years for radio leakage from an extraterrestrial civilization or for signals intentionally beamed toward us.

SETI is far more than a single effort. Like the voyages of exploration that discovered the New World or the present missions of planetary exploration, the search would involve many distinct projects with definite goals in mind. These would initially be carried out along with other astronomical investigations, but a time would come when dedicated facilities would be needed. It is estimated that a facility consisting of the collecting area equivalent to a few 100-m radio telescopes and associated data-processing equipment could carry out the initial phases of the search. In the event of a positive result or strong prospects for a positive result, a more ambitious program for SETI would be needed.

There are those who believe that we should not even try to contact extraterrestrial intelligent civilizations. In fact, in 2010, one of the premier physicists of the world, Stephen Hawking (b. 1942) elaborated upon this idea. He expects extraterrestrials to be malevolent, like Columbus, and so we should not send any signal alerting extraterrestrials to our existence. However, a group of scholars (including this author), led by Douglas Vakoch (b. 1961) of the SETI Institute, has vehemently denied the Hawking hypothesis that extraterrestrials would be malevolent and have recently published a volume on what is called extraterrestrial altruism. In the end, the point may be quite moot as radio signals from this planet have been travelling into interstellar space for some time now, at least 100 years.

13.3. TRAVELING THE UNIVERSE?

The next stage after communication, perhaps carried on simultaneously, would be manned interplanetary flight, that is, travel within the Solar System. The stars are so far from us and the technological and biological difficulties in traveling to them are so great that a trip even to the closest star seems hopeless now. Just consider the time it would take, beginning with a modest round trip to Alpha Centauri, our nearest neighbor: The distance is 4.3 light years; our ship's speed is constant at 50 km/s, slightly more than we need to escape from the Solar System, at Earth's distance from the Sun. Our round trip will take about 52,000 years.

If this kind of time scale for space travel is unacceptable, then how then do we get around time? The alternative is to use a ship that can move at something approaching the velocity of light, say 95 percent of it. This will certainly shorten the trip and also let us take advantage of Einstein's relativistic time dilation. The implications of time dilation for space travel can be illustrated by the following example.

Astronaut A leaves the Earth on a round-trip flight to some star 12 light years away, at a speed of 60% the velocity of light. At the same time, astronaut B is to travel in the opposite direction on a round-trip voyage to another star, also 12 light years away, at a speed of 80% the velocity of light. A third person, C, will remain on Earth to monitor their flights. Before takeoff, the three individuals synchronize their clocks. To avoid complications in recording the traveler's clock times, we assume that the periods of acceleration at the beginning of the outward and return voyages and the periods of deceleration in approaching each star or Earth are extremely brief compared with the time spent in moving at constant velocity. These brief accelerations may therefore be neglected.

It takes a light ray 24 years to make the round trip between Earth and either star S or T. A, traveling at 0.6c, will make the trip in 24 years/0.6 = 40 years, judged by C's clock, whereas B, traveling at 0.8c, completes the trip in 24 years/0.8 = 30 years, again according to C's clock. But A's clock runs slow by the contraction factor $1 - (0.6)2 = 0.8$, so that A's round-trip time will be 0.8(40 years) = 32 years, according to A's clock. B's clock also runs slow by the contraction factor $1 - (0.8)2 = 0.6$, and hence B's time will be 0.6(30 years) = 18 years, according to B's clock.

Suppose all three individuals are 20 years old at the start. When B returns, she finds that C is 20 + 30 = 50 years old, and she is 20 + 18 = 38 years old. When A returns, C is 20 + 40 = 60 years old, and A is 20 + 32 = 52 years old. When A and B meet again on Earth 40 years later, A will be only 4 years older than B because B (38 + 10 = 48) came back sooner than A did by 10 Earth years. Because biological aging includes a measure of time in molecular cell growth, we presume the ages here are biologically correct.

The recipe for living forever is not simply to move in one direction at a speed close to the velocity of light. An observer who leaves an inertial frame of reference must return to it in order to collect the benefits of the fact that his or her time lags behind that of the observer who remains in the inertial frame. And while in the new frame of reference the biological clock of the body may run slower, but so do all bodily activities, so that the pace of living is perceived to be the same.

From the results of the preceding

example we can refigure the trip to Alpha Centauri discussed earlier. If we could accelerate the spaceship continuously at a constant rate of 1g, that is, equivalent to the acceleration of gravity that we experience on Earth's surface, we should feel no great discomfort. The best technique then would be to reverse the acceleration halfway out and come to rest in the vicinity of Alpha Centauri and then on the return trip accelerate at 1g to the halfway point and decelerate at 1g the rest of the way to Earth. Taking acceleration and deceleration into account, the round trip by Earth clocks would be 12 years, but the contracted time aboard the spacecraft, with a maximum speed of 95 percent of the velocity of light, would be 7.2 years.

Thanks to Einstein's Relativity Theory, a journey to the Andromeda galaxy could theoretically be made during an astronaut's life span. Unfortunately, the power a spacecraft would need for an even less ambitious interstellar flight is overwhelmingly large.

On June 24, 1947, near Mount Rainier, Washington, a salesman named Kenneth

Arnold, while flying his own plane, claimed to have seen nine crescent-shaped disks flying near the mountain. His report opened the modern era of "flying saucer," or unidentified flying object (UFO), reports. Arnold's report was neither the first nor the last in that year, but it is the one that caught the attention of the news services. Since that news event in 1947, there have been thousands more in the world press and something in excess of 100,000 UFO reports that never became major news events.

In reality, the UFO "phenomenon" consists of reports of sightings and not the sightings themselves, since for most sightings there is no way of knowing what was observed, if anything. Many sightings, even those with "photographic evidence," have later been admitted to be hoaxes. Although the majority of sightings are by honest and sincere persons, where large volumes of reports have been studied, about 95 percent have been easily attributed to misidentification of "natural" phenomena. This includes airplanes, weather balloons, artificial satellites, planets (Venus has prompted more reports than any other cause), bright stars, meteors, ball lightning, flocks of birds, clouds, reflected lights, and luminous insects.

As for the reports that cannot be readily attributed to natural phenomena, the degree of strangeness in a report seems to be inversely correlated with its credibility. That is, where the strangeness is high, such as supposed extraterrestrials waving at the observer, the credibility is low. For example, only one witness saw the event or lighting conditions were extremely bad. There are virtually no high-strangeness, high-credibility reports.

If Earth is not now being visited by extraterrestrials, is it possible that extraterrestrial explorers happened this way in the past and left some sign of their visit? The sign could be some inanimate object or marking, interaction with primitive peoples if they existed, or even we ourselves if the visitors decided to seed the Earth with intelligent life. Yes, this is a possibility. Then, are any of the purported signs reasonable evidence for a visit in the past by extraterrestrial intelligence? They are probably not. For example, the "pictures of space gods," from murals, rock paintings, and pottery figures purported to show extraterrestrial visitors are indeed strange by today's standards. But the anthropologic picture of the culture in question can and does provide a cultural context for the putative space god, so that it is neither necessary nor reasonable to resort to an extraterrestrial visitor as an explanation.

Although it may seem that we are unduly harsh on the possibility of a visit by extraterrestrial beings, this is not truly the case. We find nothing that says that a visit is impossible, in the past, now, or in the future; but what we find woefully unconvincing is the supposed evidence that it is occurring now or has occurred in the past. If it has not already happened, is contact between us and other intelligent beings possible in the future? If they exist, we think it possible at some point. However, there are some very large ifs between now and that first meeting.

For the sake of discussion, assume that it will happen. Then will it be a chance meeting or prearranged? We find it hard to believe that intelligent beings will simply drop by. Instead, it is more probable that the encounter will arrive through a long sequence of message exchanges and that we shall have been learning from those beings for a very long time before the face-to-face meeting.

Chapter 14
Solar System Revelations

The place to end our discussions of the Solar System is, once again, with us. Our species, Homo sapiens, actually began to explore the Solar System some 60,000 years ago, when our ancestors first began the exploration of this planet, coming out of Africa. Homo sapiens migrated to Europe and Asia, then to the Americas and the South Pacific. Our species, in the form of Robert Peary (1856-1920) and Matthew Henson (1866-1955), first reached the North Pole in 1909. The South Pole was reached in 1911 by Roald Amundsen (1872-1928). Today, all over the surface of the Earth, Homo sapiens number over 7 billion.

14.1. SOLAR SYSTEM EXPLORATION

From the surface of the Earth our species next began to explore what lies beyond. We may consider our exploration of space to actually have begun in 1783 when Joseph Michel Montgolfier (1740-1810) and his brother Jacque Etienne Montgolfier

(1745-1799) took the first recorded hot-air balloon journey over Paris, France. With the first sustained controlled flight of an airplane in 1903 by Orville Wright (1871-1948) and his brother Wilbur Wright (1867-1912), the exploration of near space continued and flourished further.

The exploration of space with rockets blossomed with Robert Goddard (1882-1945) and Wernher von Braun (1912-1977). The first satellite to orbit the Earth was launched by Russia in 1957. It started a "space race" which led to the first members of our species to walk on the surface of the Moon, in the form of Neil Armstrong (1930-2012) and Edwin "Buzz" Aldrin (b. 1930).

Summarizing our exploration of the Solar System, we find that there have been some 72 missions to the Moon, our nearest sizeable cosmic neighbor. There have been two missions to Mercury and some 40 missions to Venus. Mars has been the object of study for some 38 missions. Jupiter has seen some 8 missions, with Juno being the most recent on its way to an encounter in 2016. Saturn has been the target for 5 missions, including a probe that landed on the surface of its moon Titan, the only satellite in our Solar System with a substantial atmosphere. Uranus and Neptune have only been visited by human built probes once, in the guise of Voyager 2. Pluto is yet to be visited by the New Horizons mission, set to visit Pluto in the summer of 2015.

As of now, there are only five spacecraft built by our species that will ever leave the Solar System. These are the Pioneer 10, Pioneer 11, Voyager 1, Voyager 2, and New Horizons spacecraft. Voyager 1 is currently the most distant, and if it has not already, it will be the first of these to leave the confines of our Sun and enter true interstellar space. Will humans ever again venture off this planet and continue our exploration of the Solar System with our own beings? Finally, who reading this book will be able to say that a spacecraft containing some samples of our species will be heading to another star system and continue human exploration somewhere beyond the Solar System?

www.ingramcontent.com/pod-product-compliance
Lightning Source LLC
Chambersburg PA
CBHW031812170526
45157CB00001B/32